The Fatal Species

In honor of <u>Noam Chomsky</u>

Andrew Y. Glikson

The Fatal Species

From Warlike Primates to Planetary
Mass Extinction

Andrew Y. Glikson
Earth and Climate Scientist
School of Biological, Earth
and Environmental Science
The University of New South Wales
Kensington, Australia

ISBN 978-3-030-75470-9   ISBN 978-3-030-75468-6   (eBook)
https://doi.org/10.1007/978-3-030-75468-6

This Springer imprint is published by the registered company Springer Nature Switzerland AG
The registered company address is: Gewerbestrasse 11, 6330 Cham, Switzerland

Preface

Are we fit to defy the blueprint of our evolution? In a kaleidoscopic journey from the prehistoric myths of ancient Australians to the 'moron-dominated global disinformation jungle of today', Glikson shows us that we are foundering on our path. Swarm intelligence has turned us into a super-species out of control where leaving what is left of our planet could one day be humanities only chance of survival. This lavishly illustrated little book is a blend of science, history, philosophy, art, and poetry. Its carefully crafted prose grips the reader from the beginning and does not let go. It is certainly provocative, but it does not preach. Rather, it reaches out to our humanity and will be fondly remembered for doing so. There are obviously hundreds of publications about any of the main aspects of this book, but I know of none that go anyway near its extraordinary blend of human endeavors. *Professor John Veron*, Australian Institute of Marine Science and International Society for Reef Studies

"The book is much more. It bowled me over with its depth of scholarship, clear writing and beautiful illustrations to provide insight into the working of the collective human mind by bringing together evidence from archaeology, literature, the arts, the sciences, animal and human behaviour to illustrate our current predicament. It leaves me searching in my own mind to equate humanity's current indolence to the threats confronting us, in stark contrast to our ability for space travel, for the creation of computers which can 'think' thousands of times faster than we can and the intellectual Emeritus Professor of Medicine at Adelaide University ability to write books like The Fatal Species." *Professor David Shearman* Emeritus Professor of Medicine at Adelaide University.

This wonderful book by Andrew Glikson says it all!!! *Dr Helen Caldicott*. Australian physician, author, and anti-nuclear advocate.

A remarkable blend of sobering science, engaging visuals and poignant bits of literary awareness. The course of human evolution on Earth has delivered us to our darkest reckoning. The Fatal Species is an elegy for an intelligent civilization struggling to get past its own worst instincts. *Professor Geoffrey Holland*, Millennium Alliance for Humanity and the Biosphere, Stanford University.

I quite like the manuscript. It is certainly different as it progresses rapidly through a range of ideas and images and finishes with a surrealist rather poetic personal summary of the human condition. It raises lots of questions and issues in ways I have not seen elsewhere. For instance the cooperative behaviour of ants suggesting some collective form of intelligence is not necessarily original but a fun idea that fits well within the context. It could be mentioned that ants actually go to war. I like the way the book presents a different way of describing the human condition through history to an uncertain future. I find it to be quite readable even though the peppering of ideas is a little scatter-gunned. Half prose half poetry. An interesting mixture, it does project a sense of mystery. *Professor Bob Pidgeon, Curtin University*.

Canberra, Australia   A/Prof. Andrew Y. Glikson

Acknowledgements

I am obliged to Brenda McAvoy for meticulous editing and proof reading of the book and for numerous discussions. Professors John (Charley) Veron, David Shearman, Geoffrey Holland, Bob Pidgeon and Dr Helen Caldicott provided generous reviews of the manuscript. Professor Will Steffen offered helpful comments on climate science aspects of the book. I thank Noam Chomsky and Helen Caldicott for correspondence and Judith Crispin for the citation.

Preamble: The killing of Gaia

Humpty Dumpty sat on a wall
Humpty Dumpty had a big fall
All the king's horses and all the king's men
Could not put Humpty Dumpty together again

History is nearing its nadir where a species of warlike primates is destroying the delicate web of life perceived by Charles Darwin (1859) in The Origin of Species, committing the fastest mass extinction in the history of nature, with global temperatures incinerating the biosphere by several degrees Celsius within a lifetime. From the blood drain rituals of the Maya and Aztecs, to the gas chambers of Auschwitz, to greenhouse gas saturation of the atmosphere, repeated generational sacrifice manifest dark demons and misogynous violence lurking inside the human psyche. With full knowledge Homo "sapiens" is proceeding to transfer every accessible molecule of carbon from the Earth crust to the atmosphere and hydrosphere, including extensive perforation of the crust allowing the toxic gases of fossil early biospheres to escape, saturating the atmosphere in an auto-da-fe of the terrestrial biosphere ensues, burning the forests, acidifying the rising oceans, flooding the cradles of civilization in the great river valleys—the Nile, Mesopotamia, the Hindus, Ganges, Mekong, Yellow River, Po and Rhine Rivers. As amplifying feedbacks to global warming—fires, methane leaks, ice melt and warming oceans—intensify at a pace exceeding any recorded in the geological past, societies are pouring their remaining resources into preparations for wars. These include nuclear wars, whose probability increases with time, triggered by arsenals many thousands of missiles strong, posing a fatal threat to human existence as well as other species. Humans, having mastered fire, bestowed by technical brilliance and artistic excellence, have emerged in the last interglacial as civilizations perpetrating major bloodsheds called "war". Long suffering from illusions of omnipotence and omniscience, paranoid fears, a warlike mindset, aggression toward the animals and disrespect of females, humans are embarking on a war against nature, culminating the absurd conflict between the mind and the heart of the species. Orwell's 1984 is already here when a moron ruler is trying to redefine reality. Eli is resigned to passing away along with nature. Existentialist philosophy may allow humans a degree of solace, as expressed by Judith Crispin: *"It is the transitory nature of things that makes them beautiful. The only things that don't change are dead things. But it is difficult to let go of the idea that things can be permanent – especially when it is the world we are speaking about. The silver lining is that the world seems so much more beautiful when we know that we are saying goodbye to it, at least to the way we know it. Time and time again he returns to the old ideas of Grace and Hope. We hope for a miracle, but we accept the non-appearance of a miracle with grace".*

Contents

Rare Earth

1

History is a nightmare from which I am trying to awake.
(James Joyce)

Unique among dead crater-scarred planets the Earth stands out—a fragile jewel coated by azure blue oceans ringing orange-brown land masses, shrouded by feathery clouds (Fig. 1.1). Life, originating from carbon compounds, as well as extinguished by excess organic carbon, has emerged around thermal springs through unaerobic reactions and through oxygen-releasing photosynthesis, at least but likely much before 3.7 billion years ago, when the planet became a haven for incipient life forms, including bacteria and stromatolites (Avramik 1992) (Fig. 1.2a), generating local oxygenated atmospheric layers. Evolving through alternating tropical and glacial eras to transient life-rich periods, surviving major mass extinctions and culminating in a biosphere teeming with migrating herds, whales, turtles, arthropods, flocks of birds, the planet reached a point when a species of primates is transferring hundreds billions tons of carbon from the Earth crust to the atmosphere.

While the enormity of the shift in state of the terrestrial climate renders it almost a taboo subject, the greenhouse gas-heated atmosphere—the lungs of the Earth—is inexorably leading to a major mass extinction. As these lines are written only those blissfully unaware of the evidence for the progressive inhabitability of the Earth (Wallace-Wells 2019) can remain optimistic. The species, consisting of a multitude of innocuous individuals is disintegrating into warring tribes dominated by a nefarious swarm mentality, is overpowering the habitability of the biosphere.

Fig. 1.1 Earth (NASA)

Humans share the Earth with a multitude of intelligent species, each special in its own right, but a self-serving double standard directed toward animals, birds and insects, assumes as if humans are "*thinking*" while other species act through "*instinct*". However birds excel in global migratory and navigational skills based on geography, the star and magnetic orientation, as does the albatross. Birds master the arts of home making, offspring rearing, communications, fire foraging, displaying intelligence as high as pre-historic human clans. Termites, bees and other arthropods are able to construct elaborate cities, including nurseries, granaries and royal chambers, expressing social organization no less advanced than human civilizations.

The appearance of creatures that learnt how to kindle fire has meant the genus *Homo* could harness energy orders of magnitude higher than their physical ability. Large parts of the flammable carbon-rich biosphere could be extracted and ignited, allowing humans to release copious amounts of energy (Fig. 1.2b). Given "*too much power*", while most individuals remain kind and generous, once large populations combine human colonies can become a monstrous force expressed in conflicts and wars.

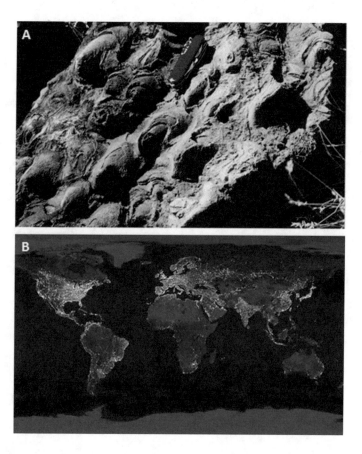

Fig. 1.2 Life from a 3.43 billion years-old Stromatolites (Pilbara, Western Australia), to b the hydrocarbons-lighted Earth (NASA)

Ancient Fires

2

Hominids are *Earthlings*, springing from the Earth to which they are connected through sight, touch, scent, intuition, orientation and, originally a deep sense of reverence toward its creatures. Prehistoric humans foraged, hunted, nurtured, fought and killed, with relics still surviving in remote parts of Earth, including the Australian aborigines:

Two million years ago the gathering thunderstorm finds Juti separated from his clan, lying in the bush, cold and hungry. A lightning strikes a nearby tree, sparking a fire. The glow warms Juti's naked body. Late into the night, as the flames ebb, Juti picks a smoldering twig and lights a small fire. By morning the flames die, he pokes a stick into the ashes. A thin pall of smoke rises. Homo Prometheus is born (Figs. 2.1 and 2.2).

Annumarrupitti Spirits

In your eyes Marlee
I see the tribal hunter
Wandering away from the kapi
To spear a Kanula
In faraway brown hills

In your voice Alinta
I hear the old tjilpi
Handing down ritual law
Master of ancient craft
Carving a wooden tinka

In your smile Pala
I feel the innocent goodness
A tradition of sharing
Amongst vanishing tribes
As your dark face glows.

A. Y. Glikson, *The Fatal Species*,
https://doi.org/10.1007/978-3-030-75468-6_2

Fig. 2.1 Fire on the rocks. A mosaic inspired by a painting by Ainslie Roberts '*The Origin of Fire*'. By Miryam Glikson, with permission

Fig. 2.2 Bradshaw rock paintings, Kimberley region of Western Australia (Creative Commons)

In your words Jeda
I perceive a temptation
The lure of the city

Tearing you away
From the old tribal way

Children of the desert
Human torch bearers
Overwhelmed by the crazed age
Of computer-controlled automatons
Such as me

Standing there in the distance
I barely could see you
A shapeless dark lump
Crouched in the dust
Blended with the ground
Indistinct from the Earth

But when I came near
Your head hardly rose
A tiny bone-dry woman
Clad in soiled rags
Betraying a slight tremor
A vibe of the Earth

Suddenly Ezekiel's dream resurrected
When you spring back to life
Deep brown eyes sparkle
A muffled voice mumbles
In a strange native tongue
An echo of Earth

An orange moon again ascends
High above Bell Rock, surging
To survey barren tribal lands,
Where wreathed men are emerging
In a corroboree dance

Soft red flames glow among the dunes
As elders gently chant
Ancient Centralian muted tunes
On humans and green ants
Shared with boulders and trees

Then my soul's fever soars
In search of a meaning
Of time forgotten ancient lore
While seconds trickle, singing
Life's sand pouring out

What shall I do with love
Humans cannot face
Perched in glass cages
Fleeing in terror from traces
Of spirits that inhabit Earth.

by Andrew Glikson

Bell Rock

For days Eli kept looking for a rock sample for age determination of the intrusive Giles Complex in central Australia, consisting of magnesium and iron rich igneous rocks derived from deep inside the Earth. Approaching Annumarrupitti Eli is stumbling among black boulders of ancient lava when he notices a round white pebble which looks like a fragment of a granophyre consisting of silica and alumina-rich differentiate of the original magma. Lifting the rock it dawned on him the area is part of ancient burial grounds of the aboriginal tribe. He looked back at his aboriginal guide to find out his reaction. There is none. Tommy remains standing silently in the distance. Not receiving a nod of approval Eli decided to leave the rocks in the ground, starting to walk away with regret.

Walking about one hundred feet away from the site Eli hears a whisper behind his back. Turning around he sees Tommy holding the white rock as if it was a child. Extending his hands toward Eli Tommy mumbles: "for you".

Eli declines, saying: "Your ancestors".

But Tommy keeps extending his hands, offering the rock.

Eli feels uncertain.

Reaching for the next rock ledge he looks back. He sees Tommy placing the rock gently in the ground, patting the rock, flattening the sand around, chanting gently as in a burial ceremony.

Eli feels a deep sense of remorse.

Bell Rock, 1991.

Swarm Intelligence

Examples of sophisticated communications among species include the bee dance, bird songs, echo sounds of whales and dolphins, possibly not less complex than the languages of prehistoric humans (Vince 2014). The appearance of a species which learnt to master fire represents the evolution of swarm intelligence to a level allowing it to exploit technical levels higher than known in any other species. The swarm's cognitive abilities arise from interactions amongst separate entities within a swarm (Figs. 3.1 and 3.2), as well as between the swarm and the environment. In a paper *"The ant colony as a superorganism* Wheeler (1911) regarded insect colonies of bees, wasps, ants and termites, some of which evolved at least since the Triassic (Schmidt et al. 2012), as superorganisms controlled by minds centered on a queen.

While the brain of a termite, an ant or a fruit fly (Fig. 3.3) is orders of magnitude smaller than that of a mammal, the collective intelligence of insects allows the swarm to undertake the most elaborate reproductive strategies, construct complex nest structures, organize systematic foraging expeditions, cultivate plants and exploit other life forms such as Aphids.

Ant colonies bear close social analogies in many respects with human societies. Whereas individual ants possess a tiny brain containing 250,000 brain neurons, compared to 86×10^9 human brain neurons, they demonstrate intelligent behavior. Collectively, the entire colony has a brain as capable in reaching complex individual and collective tasks as a single human. Due to their elaborate interconnectedness with the queen and each other through chemical signals, individual members of the colony may be compared to cells working in unison within a mammalian body. According to Gillooly et al. (2010) *"basic features of whole colony physiology and life history follow virtually the same size-dependencies as unitary organisms when a colony's mass is the summed mass of individuals…these results are evidence, not only for the superorganism hypothesis, but also for colony level selection…"*. An example for the application of superior techniques among the

A. Y. Glikson, *The Fatal Species*,
https://doi.org/10.1007/978-3-030-75468-6_3

Fig. 3.1 A flock of auklets exhibit swarm behavior (Public Domain)

Fig. 3.2 A "tornado" of schooling barracudas at Sanganeb Reef, Sudan (Creative Commons)

Arthropods is the ant *Cataglyphis bombycinus* conducting solar scans enabling navigation. Who can say whether the ant operates solely by instinct devoid of a thought process? Remarkable analogies between the behaviour of Arthropods and human societies are manifest in terms of teamwork, hunting, fighting, migration and sun-guided navigation.

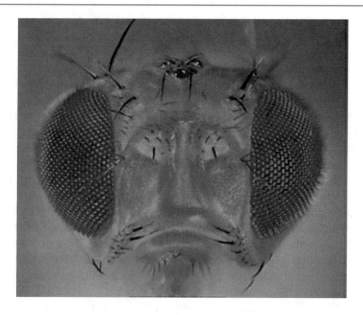

Fig. 3.3 Intelligence is everywhere. The human-like face of a fruit fly. Nature Picture Library. Fruit flies (Drosophila melanogaster) have a kind of mind. Credit Solvin Zankl/naturepl.com By Permission

Examples of super-organisms are termite colonies, the most successful insects, which colonize all landmasses except for Antarctica, building compartmentalized structures of remarkable dimension and complexity, structured to allow ventilation of the nest and analogous to anthropogenic structures in terms of the comparative proportions between individuals and the structure. Termites, dating back to the Permian, whose cockroach ancestors appeared in the Carboniferous more than 300 million years ago, have similar radiation resistance and survived extreme climates and several mass extinction events (Turner 2011). Originally living in the rainforest, fungus farming allowed termites to colonize the African savannah and other new environments, eventually expanding into Asia (Roberts et al. 2016). The termite queen is fertilized by a smaller male king, is fed by nursing ants and produces millions of eggs in its lifetime.

The nature of the collective intelligence of ants has been studied in detail, for example in Stanford University, the observations lending support to the Aristotle's dictum "*The whole is greater than the sum of its parts*". The mm-scale and complexity of insects' sensory and communicative capacities may be compared to the computing elements of microchips in terms of their high processing power. In his book "*The Soul of the White Ant*" Eugene Marais (1925) considers the analogies between communal processes within a termite society and the workings of the human body. Marais regarded red and white soldiers as analogous to blood cells, the queen as the brain and the termites' mating flight where individuals produce new colonies as equivalent to the movement of sperm and ova. The colony super-organism possesses cognitive powers allowing it to respond adaptively to environmental conditions beyond the ability of individual termites (Bonabeau et al. 1999).

Based on detailed observations of termites in Namibia Scott Turner (2011) states *"Eventually it hit me: these are not Robots but are living things with individuality, wants and desires. A robot cannot ever "want" to be groomed or "want" a drink, or "want" to give water to another, but termites seemingly do. This gives termites, individually and collectively, something like a* soul*—an animating principle that one does not find in machines. It needn't be some vital "stuff" as the ancients once thought, but still something ineffable that makes life distinct from nonlife."*

The swarm's cognitive abilities arise from both interactions amongst the individual agents within a swarm as well as interaction of the swarm with the environment. The social structure of termite, ant and wasp colonies has been likened to a superorganism, in which the colony has many of the attributes of a more conventionally defined organism, including physiological and structural differentiation and coordinated and goal-directed action. In so far as cognition constitutes a social phenomenon insect colonies should be imbued with cognition, whether units of the social system are cells, or neurons in brains, as among mammals, or organisms within superorganisms, as are colonial insects (Gillooly et al. 2010; Rettner 2010; Turner 2011).

Whereas the brain of a termite or an ant is smaller by orders of magnitude compared to the human brain, where ants construct elaborate nests, undertake a foraging expedition, cultivate Aphids, with analogies to Neolithic farmers, who is to say they operate only by "instinct" while humans operate by "thought"? Who is to say their thought processes, at least as a swarm, are "inferior" as compared, for example, to those of *Homo habilis*?

The intelligent expressions and art of birds, distant descendants of the famed dinosaurs, in home making, offspring rearing, communication with humans, are hardly surpassed. In the Northern Territory, Australia, Fire foraging raptors burn bush areas to expose their prey, black kites and brown falcons pick small sticks to start a new fire. Expressions of artistic creativity testify to the imaginative faculties of birds, such as manifested by nest decoration by New Guinea Bowerbirds.

Apart from *sapiens* only few creatures are known to own an aesthetic disposition. Bowerbirds (*family Ptilonorhynchidae*) appear to entertain an aesthetic sense of beauty, decorating their nests to attract the female. Whether Birds of Paradise possess a concept of the extravagant beauty of their dance remains unknown. For humans beauty constitutes an essential foundation of visual art and music. It is an inspiring dimension arising from an aesthetic sentiment rather than logic, and at the same time subjective and closely intertwined with culture. Not every Inuit would be inspired by a painting by Rembrandt and not every aboriginal by a nocturne by Frederick Chopin, although likely most humans would love the colors of the bird of paradise and a sunset over the mountains.

Extreme levels of individual intelligence, family cohesion, social intelligence and swarm organization are displayed by bird species, notably where albatross and penguin pairs nurture their chicks, or separate for many months to fly over the

Fig. 3.4 Waved Albatross (Phoebastria irrorata), Espanola Island, Galapagos Islands, Ecuador (Creative Commons)

oceans to catch fish and then return to feed their partner and offspring (Fig. 3.4). Here, the combination of paired family bond, navigation skills and commitment to perpetuate the family and the species are no lesser than those of some human—arguing against an artificial anthropocentric narcissism.

Birds possess small brains (raven brain is 15 g) compared to most mammals (human brain = 1508 g), but a high ratio of brain weight to body weight underlies a high intelligence (Table 3.1). Large parrots and corvids have especially dense

Table 3.1 Brain to body mass ratio

Species	Brain:body mass ratio (E:S)[4]
Small ants	1:7
Small birds	1:12
Mouse	1:40
Human	1:40
Cat	1:100
Dog	1:125
Frog	1:172
Lion	1:550
Elephant	1:560
Horse	1:600
Shark	1:2496
Hippopotamus	1:2789

clusters of neurons in their forebrains, allowing a high cognitive ability. Among birds, those that are larger-brained (corvids, parrots and owls) are considered the most intelligent. These birds, with forebrains and neuron systems analogous to those of apes, live in complex social groups and have a long developmental period before becoming independent, possessing ape-like intelligence (Emery 2004). This author suggests complex cognition has evolved in species with very different brains through a process of convergent evolution rather than shared ancestry. Although birds do not have a neocortex, the crow family has evolved complex behavior allowing tool use, deception, face recognition and other faculties, some inherent from their dinosaur ancestors (Callaghan 2016). Corvids, including crows and ravens, use tools, solve problems, and recognize themselves in the mirror. Whereas a raven's brain is only 15 g as compared to a chimpanzee's brain of 420 g, their behavior seem inexplicably advanced for their small brain size, compensated for by the density of neurons, about twice as dense as mammals of the same size. This included expansion of the fore-brain areas known as the optic tectum and the cerebellum, which have allowed birds to escape the K-T boundary asteroid impact catastrophe.

Bar-headed geese (*Anser indicus*) soar over the Himalaya, whales and albatrosses navigate around Antarctica, sea turtles travel the oceans, penguins cross rocky coasts to reach their families, gazelles and leopards outpace humans. If there is a justification for claims of human exceptionalism it is the application of fire, enhancing the human energy output by orders of magnitude and constituting the foundation of technology, science and civilization.

Among animal brain to body weight ratios dolphins have the highest ratio of all cetaceans, second only to humans (Fig. 3.5; Table 3.1). Arthropods have a high brain to the body mass ratio (E/S), (small ants 1:7) as compared to birds (small birds 1:12), mice (1:40) and man (1:40). In terms of the number of neurons Insects have

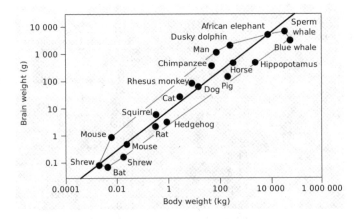

Fig. 3.5 The relations of brain weight to body weight (Creative Commons)

about 250,000 neurons compared to humans (21×10^9 neurons). In terms of swarm intelligence the collective number of neurons and thereby collective intelligence may be higher by orders of magnitude.

The culmination of natural history epitomized by *Homo sapiens*, constituting a unique development in nature, is turning out as a brief episode of mass extinction which questions the supposed superiority of *Homo Sapiens*.

Amazons and Misogynous Apes

4

Among the Arthropods it is the queen, or several queens, who dominate the colony, while the males fertilize the females and specialize in fighting. By contrast among mammals it is the males who are more commonly dominant and may be vicious towards females, as is the case with Chimpanzees, exercising forced copulation, the robbing of food and even murder (Ananthaswamy and Douglas 1998). The patriarchal behaviour of chimpanzees, who share a last common ancestor with humans until about 4 million years ago, sheds some light on the behaviour of human males who often tend to congregate separately from women, almost as if the latter constituted a separate troupe. In patriarchal societies men are more likely than women to hold social, economic and political power, perhaps in part because they are generally physically stronger than women. In circumstances where women move to live with their husband's patrilocal residence, men tend to have greater power and privilege.

By contrast to Arthropods, among many mammalian species it is the male who is dominant as the head of the tribe, chief fighter and originator of offsprings. The violence among males, expressed through spirals of killing, raping and waging wars, as documented in *"Demonic Males: Apes and the Origins of Human Violence"* (Wrangham and Peterson 1996), explains much of human relations and history in misogynous terms.

It may be tempting to assume male dominance is the natural state of human society, although this has not always been the case. Bonobo societies are both patrilocal and at the same time female-dominated, where females often have the upper hand as they cooperate and are closely associated. During most of Homo's history humans have been hunter-gatherers, where patrilocal residence is not the norm. With the advent of agriculture and homesteading around 12,000 years ago (Langergrabe et al. 2013) people began settling down and male-female living circumstances and relations may have changed. Power shifted to physically stronger males where fathers, sons, uncles and grandfathers began living near each other, property was passed down the male line and female autonomy was eroded—

A. Y. Glikson, *The Fatal Species*,
https://doi.org/10.1007/978-3-030-75468-6_4

patriarchy emerged (Ananthaswamy and Douglas 2018). Agriculture is indeed correlated with patrilocal societies. An inherent biological and psychological distinction between males and females led to sharp separation in their respective roles and positions in prehistoric and historic societies, although exceptions are known. At various stages women rebelled, attaining influence and even dominance, when they could combine their compassionate, intelligent and practical attitudes with their powerful functions in society. The legendary Amazons (Foreman 2014) are one example, albeit shrouded in mystery, perhaps because it was mainly men who wrote their stories?

Despite their common subjugation women were elevated to divine positions, for example in Middle East civilizations, including the Sumerian Akkadian, Assyrian, Babylonian goddess Inanna, Ishtar. Typically these deities are portrayed as winged fully armed fighters with spears and arrows, taming lions and birds of prey (Figs. 4.1 and 4.2).

Fig. 4.1 The goddess Ishtar or Inanna on an Akkadian Empire seal (2350–2150 BC), carrying weapons on her back, trampling a lion on a leash was one of the goddesses worshipped throughout the ancient Mediterranean and Near East referred to as the Queen of Heaven, including Anat, Isis, Ishtar and Astarte. In Greco-Roman times, Hera and Juno bore this title (Creative Commons)

Fig. 4.2 The Queen of night relief 1800–1750 B.C.E., Old Babylonian, Rectangular, baked clay relief panel, modelled in relief on the front depicting a nude female figure with tapering feathered wings and talons, standing with her legs together. Trustees of the British Museum (Creative Commons)

Ishtar—A story

The fiery desert wind pushes the dunes northward all day, their dusty crests curling toward a scorching sun, burying me and my Afghan under drifting walls of sand. At night, during lulls in the storm, we wade our way north and west through sandy waves, guided by Andromeda, Perseus, Cassiopeia, the Pleiades and Venus—queen of Pan's divine realms, goddess of love, beauty, desire, sex, fertility, prosperity and victory, torch bearer of Ishtar-Inanna and Ashtoreth. By dawn she twinkles toward me as she fades away against the crimson crescent of the eastern horizon where, once again the sun's rising furnace drives us to dig a shelter in a small clearing among the dunes.

There buried, hugged against my Afghan, I dream of a palm-fringed oasis, of marble-white figures with deep dark eyes, dancing Centaurs and nymphs, quarrelling gods on Mount Olympus, fiery sacrifices to Baal. I dream of slaves nailed to crosses strung along the via Apia, of the demise of nature's spirits, the lifeless absolutes of one god, one king, one patriarch, medieval crusades, the enslavement of women, medieval gendercide, auto-da-fe.

From the crest of the last sand dune mountain unfolds the mighty reed-flanked Tigris, snaking its way along the ruins of Nineveh as it has done for millennia. In my mind's eye I see a fifteen-gate wall surrounded by canals, a towering wing-framed statue of a woman clad with spears and arrows, her foot subduing a lion. A guiding star shines over temple processions through avenues decorated with dragons, lions and bulls. The vision disappears; from behind some ruins in a boulder strewn field appears a Bedouin, a watering jar in his hand:

"Salam alaikum. I am Masud, welcome to the graveyard of the gods, you must be thirsty?"

"Did I see you watering the stones?"

"A tribute to the gods", he mumbles, "at night their spirits visit the stones, they descend into the graves, they rise, then disappear".

"Why, mortal gods?"

"Mortal only in body, their spirits live forever"

"And you?"

"I keep the tombs alive, by the grace of Ishtar, hers and those of her kin".

The Bedouin points to a woman-shaped boulder surrounded by smaller rocks.

"Can we set camp here for the night?"

As she crouches to the ground, my Afghan releases a gurgling belch, soon bright flames dance on a small clearing opposite the boulders. The glow of the Pole star above the northern horizon fuses with dancing flames of the camp fire.

"Will you join us for what is left of our food?"

"Thank you, Inshallah"

"Will you tell me about Ishtar?"

"Only the one who crosses the Rub El Khali's Empty Quarter is allowed the knowledge" "How can I not honor you? El Akhbar!"

"Ishtar was the daughter of Inanna the moon god, from whom she received the divine laws, sister of Shamash the sun god, Ishtar grew in the silvery aura of her kind father's spirit, inspired by passion and ethereal devotion, driven by absolutes, fighting for love, dying for love".

"In her youth two lions came under her aura. Ishtar loved Sennacherib—a sensitive lion, but overcome by compassion she married Ashurbanipal, spending an eternity in exile. Ishtar bore demigods, to see them flap wings and fly. Left alone, her divine cravings on Earth unfulfilled, her healing spirit descended into the underworld, nurturing the ill and the dying. Yielding her gifts at every gate to finally find herself naked in front of Ereshkigal, queen of the underworld".

A full orange moon reflects in the Tigris. Massud pointed to two lion-like boulders and to four winged stones not far from the woman-like boulder. "One day Ishtar met Sannacherib in an underworld cavern. Channeling her pain she tried to resurrect her life around him, alas for by then his roots were deeply intertwined with those of his offspring.

"What happened?"

"No one knows. Some say Ishtar discovered self-love, of her own beauty and wisdom"

"What does that rolling stone, away from the rest, mean?"

"Oh, that's the stranger she met during her anguish, from nowhere, belonging nowhere, she believed he was sent by the moon god to hold her hand in the dark as Orpheus has, with a message of life, but he proved to be only human".

The wind ebbs, the eerie night air gives me no relief, a distant howl mourns the setting moon, Venus the morning star reflects dimly from the stones, the breeze returns, whispering a growing hush in the distance, suddenly the silence is broken by a flutter of wings. My raised eyes meet pale spirits hovering all over the field— white-clad women, men and children circle the stones, some sink underground, some reemerge. A gust of wind pulls me to the distant round stone, for a few moments I float effortlessly above, then my heart merges with my entombed body. Eternity glides by in the company of an invisible image, looks back then it disappears, I am drawn into a ring of spirits dancing to the tune of cymbals, singing a song of universal sisterhood, love and joy.

For a while I whirl around holding Ishtar's white hand against my chest. I look into her dark eyes, she twinkles. I wake alongside my Afghan half buried in the shifting sands under a clear sky to a ray of the morning star.

Steeped in-between evidence and legend were the Amazons, women archers on horseback (Figs. 4.3 and 4.4), extoled in fighting, hunting and sexual freedom. Battle-scarred female skeletons buried with their weapons prove they were not merely figments of the Greek imagination. According to Mayor (2014) warlike nomadic women of the Eurasian steppes inspired tales in ancient Egypt, Persia, India, Central Asia, and China. Combining classical myth and art, nomad traditions and scientific archaeology, the author reveals intimate, surprising details and original insights about the lives and legends of the Amazons. Sifting fact from fiction he shows how flesh-and-blood women of the Eurasian steppes were mythologized as equals to men.

Fig. 4.3 Wounded Amazon of the Capitoline Museums, Rome (Creative Commons)

Women in Nubia exercised significant control over their communities, manifesting the cult of Isis, as contrasted with the Egyptians' worship of Ra, the male sun god. The Ruling queens of Nubian/Kushite Empire during the golden age of the

Fig. 4.4 Woman or goddess ("La Parisienne") from the Camp-Stool fresco, c.1350 B.C.E., western wing of the palace at Knossos (Creative Commons)

Meroitic Kingdom (800 BC–300 AD) (Kneller 1993) signified the offspring of the gods and bearers of the cult of Isis. Isis became the most popular thanks to her role as the devoted, untiring, nurturer of the land and culture of Egypt and Nubia.

Several south-east Asian cultures are known where females played decisive social and tribal roles. Palawan tribe men and women in the Philippines live in perfect equality, with values of goodwill, generosity and mutual assistance (Vallombreuse 2015). This author has taken poignant photographic portraits of women full of strength, life and truth, who occupy a central place in the social and spiritual foundations. These women preserve and advocate equality between the sexes.

Fig. 4.5 Women of Archaic Athens fetching water from a fountain house, on painted terracotta Hydria (water jug) of ca. 530 BC. Attic Greek, Doris, Metropolitan Museum of Art (Creative Commons)

In the Khasi society of North-East India, a matrilineal and matrilocal culture, children bear the name of their mother. The practice of children living with the mother's family ensures the well-being of children should the parents be separated. The youngest daughter inherits the family land and properties. Women of the Musuo society in south-west China run households including several families, where the head matriarch s of each village govern the region by committee. These women have a unique status and are free to take different partners and still carry the family name. The children are raised in the mother's house with the help of her brothers and the rest of the community.

The role and status of women in Greek and Roman societies has often been obscured by the bias of male writers, history being written by "winners". Meticulous studies of women in the Classical World, Greece, Crete (Fig. 4.4a) and Rome, from slaves and prostitutes, to Athenian housewives (Fig. 4.5), to Rome's imperial families, unravel detailed evidence about their daily lives (Fantham et al. 1994).

Fig. 4.6 A scholastic woman
with wax tablets and stylus
(so-called "Sappho")
(Creative Commons)

The historical and cultural contexts are revealed through poetry, vase painting, architecture, funerary art, women's ornaments, historical epics, political speeches and ancient coins. Art and literature highlight these women's creativity, sexuality, coming of age, marriage, child rearing, religious and public roles. Apart from the legends about women like Cleopatra, Dido and Lucretia, images on urns of graceful maidens dancing remain little-interpreted. Stories of the wild behaviour of Spartan and Etruscan women, love poetry of the late Republic and Augustan age, and traces of upper- and lower-class life in Pompeii (Fig. 4.5) are preserved by the eruption of Mount Vesuvius in 79 C.E., including images of scholastic women like Sappho (Fig. 4.6).

River Empires and Divine Rulers

5

Homo sapiens emerged prior to the end of the Eemian interglacial (pre-124,000 years ago) when temperatures were rising from glacial levels by about 5 °C and sea levels rose by near-100 m Hansen et al. (2008), similar to previous Milankovitch cycles (Buis 2020). During this transition, 150 to 130 kyr ago, small clans of nomad hominids survived climate changes by migrating within and out of Africa. In the wake of climate disruptions associated with the last glacial termination about ∼ 14–7.5 kyr ago, including marked stadial cold reversals such as the *oldest, older* and *younger dryas* and *Laurentian* stadials, stabilization of the climate led to regulated river flow regimes controlled by annual climate cycles in source mountain terrains, allowing stable river terrace cultivation (Fig. 5.1). From approximately 7000 years-ago development of large-scale river irrigation networks saw the growth of Neolithic civilization along the Nile, Tigris, Euphrates, Indus Mekong and Yellow Rivers, where a rise and fall of ancient civilizations was controlled by the seasonally regulated balance between accumulation and melting of snow in their mountain sources. Seasonal regulation of river flow associated with re-deposition of fertile silt, enhancing production, was interrupted by periods of low river flow, floods and erosion of terraces.

The Tigris and Euphrates rivers, allowing irrigation from about 6000 years-ago, formed the cradle of the Mesopotamian ("Land between the rivers") civilization, where Sumerian cities between 4000 and 3100 BC were succeeded by Babylon, ruled by fighting warlords (Figs. 5.2, 5.3). During the same period cultivation along the Indus River, fed from the Himalaya, was developed by the Harappan civilization. Along the Yellow and Yangzi Rivers civilization saw the rise of the Xia, Shang and the Zhou Dynasties from about 7000 years-ago. The self-anointed rulers were supported by casts of priests who, among other, could forecast weather changes, floods and famine, and thereby food production, attaining enormous powers over the population.

The Nile River, fed by water from the Ethiopian Mountains, allowed the rise of the Old Kingdom (4660–4160 BP), Middle Kingdom (4040–3640 BP) and the New Kingdom (3550–3070 BP), with the largest pyramids built during the Old Kingdom

Fig. 5.1 An early Ramesside Period mural painting from Deir el-Medina tomb depicts an Egyptian couple harvesting crops (Creative Commons)

Fig. 5.2 Relief depicting Ashurbanipal fighting a lion with a stylus tucked into his belt, 645–635 BC (Creative Commons)

Fig. 5.3 Victory stele of Naram Sin, the Akkadian Empire in ancient Mesopotamia, 2350–2000 BC (Creative Commons)

(Table 5.1). A major ~4.2–4.0 kyr desertification (Brookfield 2010) occurred when the White Nile ceased to flow continuously, with major effects on cultivation, likely causing the collapse of the Old Kingdom in Egypt. A similar decline in river flow affected the Akkadian Empire (Weiss et al. 1993) in Mesopotamia and Harappan civilization along the Indus River. Deep sea core sediments from the Gulf of Oman testify to a several-fold increase in wind-borne aeolian components from 4025 ± 125 BP, representing development of arid conditions in the source regions of Mesopotamian rivers (Cullen et al. 2000). Radiocarbon age determinations from

Table 5.1 A summary of ancient Egyptian history

• ~20–12.5 kyr. Frozen Ethiopian Mountains; stable sediment alluviation and terrace building by a low-river flow; Hunter-gatherers
• ~12.5–8 kyr. High floods (the 'wild Nile') due to heavy precipitation in the Ethiopian Highlands; little rain along the Nile. Increased vegetation in the source terrain leads to less sedimentation and thus increased erosion of river terraces; reduction of population
• ~8–6 kyr. Seasonal climate, stabilization of the Nile and re-aggradation of alluvial terraces, allowing irrigated agriculture
• ~7.5–5.1 kyr. Pre-dynastic Egypt
• ~5.5 kyr. Retreat of the rain belt southward
• ~4.686–4.181 kyr. Old Kingdom
• ~4.2–4.0 kyr. Desertification
• ~4.0–3.7 kyr. Middle Kingdom
• ~3.570–3.069 kyr. New Kingdom
• ~3.2–2.55 kyr. Iron age cold period

Tell Leilan, northeast Syria, uncover evidence for an incipient collapse about 4170 ± 150 BP.

Paleoclimate and archaeological records demonstrate close relations between prolonged droughts and social collapse. Overpopulation, deforestation, resource depletion and warfare have reinforced social collapse (Diamond 2011). Repeated droughts likely constituted the root factors in the downfall of the Akkadian, Maya, Mochica, and Tiwanaku civilizations. Late Holocene climate perturbations involved repeated inter-annual droughts and infrequent decadal droughts deMenocal (2001) of relatively minor effect as compared to the Anthropocene global warming.

While in pre-Neolithic societies humans had to cooperate, relying on each other within clans and tribes in their struggle to survive, they were hardly equal—following those with a physical prowess and aptitude who led the hunt, reaping rewards in relation to women and proliferation of offspring. As among baboons such leaders dominated their clans, fighting off rival contenders and upstarts and leading in battle. The regulation of competition has required the development by tribal elders of codes of behaviour, becoming systems of traditions and laws to be strictly obeyed, along with customs and lore, such as the aboriginal dreaming, connecting prehistoric humans, in particular the women, with the spirits of Earth.

Pre-historic life, focused on efforts at gathering and hunting for food, developed meanings conceived around campfires. Pantheistic theology requires human, animal, bird and fish to constitute parts of one vast tapestry of creation interconnected with the ancestral spirits. The *Dreamtime* conceived by the Australian Aboriginals represents insights into the world through creation stories of the great ancestors, where laws must be strictly observed. The natural generosity of people living at the edge of existence, such as the Bedouins of the Sahara or the Inuit in the Arctic, is rarely matched by the inhabitants of wealthy civilization.

With the Neolithic, the availability of extra food has led to a diversification of tasks and privilege, a rise of the *winner takes all* mentality, competition, conflicts

Fig. 5.4 Stone head carving of Ramesses I, originally part of a statue depicting him as a scribe; on display at the Museum of Fine Arts, Boston (Creative Commons)

and wars. Resources raised by the taxation of farmers allowed smelting of metals for tools and weapons, supporting the growth of a hierarchy and a religious order, giving rise to divine rulers (Figs. 5.4, 5.5), guards, armies and warlike absolutism, accompanied by a decline of the prehistoric sense of reverence toward nature, the *Achilles Heel* of modern humans. In the contemporary world, armies of computer programmers, lawyers and economists impose a legalistic electronic language on common sense, preventing solution to the human predicament.

Ozymandias
By Percy Bysshe Shelley

I met a traveler from an antique land,
Who said—"Two vast and trunkless legs of stone
Stand in the desert. ... Near them, on the sand,
Half sunk a shattered visage lies, whose frown,
And wrinkled lip, and sneer of cold command,
Tell that its sculptor well those passions read
Which yet survive, stamped on these lifeless things,
The hand that mocked them and the heart that fed;
And on the pedestal, these words appear:
My name is Ozymandias, King of Kings;
Look on my Works, ye Mighty, and despair!
Nothing beside remains;
Round the decay
Of that colossal Wreck, boundless and bare
The lone and level sands stretch far away.

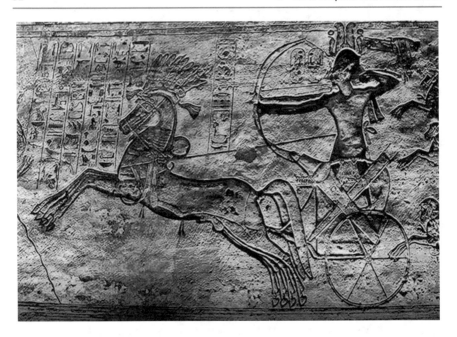

Fig. 5.5 Ramesses II fighting in a chariot at the Battle of Kadesh. Relief from Abu Simbel (Creative Commons)

The extra-wealth accumulated by cultivation of river terraces in the valleys has been spent by empires fulfilling dreams and nightmares which originated in pre-historic campfires, ensuring an immortality for the rulers, robbing and killing neighbouring people, pursuing the dream of eternal life through construction of grand monuments exploiting slave, prisoners and peasant labour, accompanied by entombed royal entourages (Fig. 5.6), grand sacrificial rituals and brutal conquests. The invention of currency and replacement of barter economy with money trans-actions have allowed transport of wealth across long distances, deserts and even-tually oceans—the loot from the subdued populations, enhancing the motivation for wars of plunder, best manifested by the history of central and South America.

By contrast to wealthy valley civilizations, traditionally mountain and desert people possessed distinct attitudes, reflecting their hardships and thereby more austere nature. An example are the Bedouins with their legendary endurance in long treks through deserts, zealous traditions, generosity and courage, impressively reported by Wilfred Thesiger in *Arabian Sands* (1959) (Fig. 5.7).

Fig. 5.6 Mausoleum of the First Qin Emperor (Creative Commons)

Fig. 5.7 Syrian bedouin, Khalil Sarkees, with family, 1893 (Creative Commons)

Human Sacrifice

Animals, while killing for food, may play with their victims, such as killer whales with pup seals, cats with mice and foxes with chickens, yet hardly go to the elaborate length humans can while torturing their victims. Mass killings are known in battles among the arthropods, for example when metabele ants invade termite nests, yet allowing the colony to survive. Grotesque forms of torture, genocide and ecocide constitute a unique human phenomenon, from tribal massacres such as of the Tutsi by the Hutu in Rwanda, to the industrial-scale death camps symbolized byAuschwitz. Recognizing their own mortality, subconsciously humans try to triumph over death by perpetrating death, constructing huge colossi for ritual mass sacrifice and gladiator games, with promised glory in the afterlife. From blood drain rituals by the Maya and Aztecs, to the gas chambers of Auschwitz, to greenhouse gas poisoning of the atmosphere, generational sacrifice manifests dark demons lurking inside the human psyche.

Through history the privilege associated with hunting and sacrifice of animals to the gods constituted the prerogative of priests and rulers, as symbolized by the Mesopotamian Epic of Gilgamesh, the demigod and his comrade Enkidu ripping out the heart of the Bull of Heaven as a gift to the sun god Shamash (Fig. 6.1). Glorious and gory rituals including claims of human lives are peppered through history. In Greek mythology King Agamemnon decides to sacrifice his daughter Iphigenia to Artemis as a payment for letting the Greek fleet sail to Troy. In the book of Genesis Abraham, about to sacrifice his son Isaac to god is halted by an angel staying his hand at the last minute.

Inanna to Anu, in the *Epic of Gilgamesh*, Tablet VI, Third Dynasty of Ur (c. 2100 BC)
Father, let me have the Bull of Heaven
To kill Gilgamesh and his city
For if you do not grant me the Bull of Heaven
I will pull down the Gates of Hell itself
Crush the doorposts and flatten the door
And I will let the dead leave

A. Y. Glikson, *The Fatal Species*,
https://doi.org/10.1007/978-3-030-75468-6_6

And let the dead roam the earth
And they shall eat the living
The dead will overwhelm all the living!

Human sacrifice, more than the stuff of legends, is repeatedly established by archaeological diggings, finding evidence of child sacrifice in many cultures, including for example sacrificial pits that dot the site of Yinxu, the last capital of the 1600 to 1000 BC Shang dynasty, the earliest Chinese dynasty of to leave an archaeological record. More than 13,000 people were sacrificed at Yinxu over a roughly 200-year period, with each sacrificial ritual claiming some 50 human victims. Wars, glorified under god or national flags, constitute the equivalent of ancient human sacrifice (Figs. 6.2, 6.3).

Fig. 6.1 Ancient Mesopotamian terracotta relief (c. 2250–1900 BC) showing Gilgamesh slaying the Bull of Heaven (Creative Commons)

Fig. 6.2 Tongue removal and live flaying of Elamite chiefs after the Battle of Ulai, at the coronation of Ummanigash and his brother Tammaritu, 653 BC. British Museum (Creative Commons)

Fig. 6.3 The Maiden. This frozen mummy was found entombed near the top of the Llullaillaco volcano in northwest Argentina. Known as the Llullaillaco Maiden, the 13-year-old was ritually killed in an Inca rite hundreds of years ago (Creative Commons)

Slavery reached a peak during the sixteenth and eighteenth centuries, constituting an exploitative, inhumane, racially founded institution rather than anything that had previously existed in Europe, Africa, or the Americas (Guesco 2017). Indigenous peoples were preyed upon for their labour in the Caribbean and throughout the Americas, especially in the silver mines of Mexico and the Andes. The deaths of Indians from epidemic diseases led to a predatory system of slavery that relied on captured African labourers. Extensive transatlantic slave trade in the seventeenth and eighteenth centuries was destined to cut cane and tend sugar mills in Brazil and the West Indies, work the tobacco fields and rice paddies of North America. Crammed at the bottom of wooden ships the suffering and death of multitudes of men, women and children bear few parallels in history.

Femicide, common during the Middle Ages, is estimated to have claimed between 1560 and 1630 the lives of around 40,000–50,000 women by hanging, drowning and burning for witchcraft (Fig. 6.4) in Europe and the American colonies. Prosecutors in the American colonies generally preferred hanging. In the sixteenth century women were burnt in both Catholic and Protestant witch trials, with the largest number between 1560 and 1630. Widespread mass witch trials occurred in the regions of the Catholic Prince Bishops in Southern Germany,

Fig. 6.4 The burning of three witches in Baden, Switzerland (1585), by Johann Jakob Wick (Creative Commons)

involving hundreds of including the Trier witch trials (1581–1593), the Fulda witch trials (1603–1606), the Basque witch trials (1609–1611), the Würzburg witch trial (1626–1631) and the Bamberg witch trials (1626–1631). In remote parts of Europe and in North America witch burning prevailed in the 17th-century, including the Salzburg witch trials where numerous children were tortured and killed, the Swedish Torsåker witch trials, and the Salem witch trials in New England in 1692.

In purporting to expose demons in the hearts of innocent victims, the witch trials betrayed the depravity of the persecutors, the judges and those in society at large who supported them, revealing barbarous forces underlying the principal institutions of society, primarily those claiming to represent the humanity of Christ. The betrayal is continuing through recent history, not least during Pius XII affiliation, described as "a fellow traveller of the Nazis".

It is hard to find a parallel through human history to the systematic indoctrination of an entire generation with a murderous philosophy, translated into genocide, such as achieved by the Nazis, converting millions of youth into a mindless almost monolithic mass murder machine, and this in one of the most civilized nation that ever existed in Europe. The temporary success of the Nazis in recruiting an entire nation to their criminal organisation poses a warning to humanity that this can happen again and anywhere (Fig. 6.5).

The ideology of Nazism has and continues to serve as a flag of convenience for killers to justify the worst atrocities, including physical and sexual violence against minorities and women. Estimates by the World Health Organization indicate that some 35% of women worldwide have experienced partner violence in their lifetime and near 38% of murders of women are committed by intimate partners, including

Fig. 6.5 Young girls cheering Hitler

intergenerational violence. Perpetrators of violence commonly have a background of child maltreatment, domestic violence against their mothers, a belief in entitlement over women, drug abuse and exposure to a culture of violence. In Australia approximately one woman is murdered every week. A 2016 Personal Survey indicates some 2.2 million Australians have experienced physical and/or sexual violence from a partner and 3.6 million have experienced emotional abuse from a partner, misogyny remaining a nasty force in societies.

On Slavery and Genocide

7

Parasitic exploitation and symbiotic relations, common among the Arthropods with an example of slave raids of Amazon ants (Polyergus rufescens) are dominant among human civilizations, with feudal hierarchies taxing the peasantry, enslaving people (Figs. 7.1, 7.2, 7.3) or driving people into abject poverty (Fig. 7.4), a process culminating in genocide (Figs. 7.5, 7.6).

Under terror most people will have no alternative but succumb to the banality of evil, the essence of totalitarian regimes being to render large populations mere cogs in dehumanized systems destined to genocide (Figs. 7.5, 7.6). In her essay Eichmann in Jerusalem Hanna Arendt (1963) states *"No communication was possible with him, not because he lied but because he was surrounded by the most reliable of all safeguards against the words and the presence of others, and hence against reality—proof against reason and argument and information and insight of any kind"* ... *"It is indeed my opinion now that evil is never "radical," that it is only extreme, and that it possesses neither depth nor any demonic dimension. It can overgrow and lay waste the whole world precisely because it spreads like a fungus on the surface. It is "thought-defying," as I said, because thought tries to reach some depth, to go to the roots, and the moment it concerns itself with evil, it is frustrated because there is nothing. That is its "banality." Only the good has depth that can be radical."*

It becomes clear language is insufficient to portray genocide. *"Then for the first time we became aware that our language lacks words to express this offense, the demolition of a man ... We had reached the bottom. It is not possible to sink lower than this ... Nothing belongs to us anymore: they have taken away our clothes, our shoes, even our hair... They will even take away our name..."* (Primo Levi).

© The Author(s), under exclusive license to Springer Nature Switzerland AG 2021

A. Y. Glikson, *The Fatal Species*,

https://doi.org/10.1007/978-3-030-75468-6_7

Fig. 7.1 A small bronze statue dating back nearly 2000 years may be that of a female gladiator
(Creative Commons)

The oblivious holocaust thinking is symbolized by the missile launch machine,
as shallow and meaningless as the Auschwitz death factory, with the gas chambers
replaced by missile fleets threatening to turn the entire planet into an open-air oven.
Fundamental philosophical notions underlying such events may be found in the
Hindu Vishnu–Brahma–Shiva trilogy and the Yin–Yang polarity (Fig. 7.7).

In a duality, symbolized by the Yin and the Yang of ancient Chinese philosophy,
opposite forces are seen as interconnected and counterbalancing in a dynamic
system where the whole is greater than the sum of the parts. Duality is inherent

Fig. 7.2 Slaves: Tarnished steel sculptures of black men and women, chained, kneeling, screaming, pleading, rust running down their naked bodies, mark the entrance. National Memorial for Peace and Justice, Montgomery, Alabama (Creative Commons)

Fig. 7.3 Sarah "Sally" Hemings (c. 1773–1835) was an enslaved woman of mixed race owned by President Thomas Jefferson (Creative Commons)

where anything and everything hold opposing realities, all of which have validity. Duality encompasses subatomic wave-particle pairs, positive versus negative energy states, electron-proton pairs, matter versus antimatter, gravity versus anti-gravity, light versus dark, conflicting human behaviour (Arthur Koestler, 1978: *Janus: A Summing Up*), where "*the human brain has developed a terrible*

Fig. 7.4 Extreme poverty. Children of road workers near Rishikesh, India (Creative Commons)

biological flaw, such that it is working now against the survival of the race. Something has "snapped" inside the brain. It is no longer necessarily a function which will lead us to a better world, but something demonic, possessed, perhaps even evil".

Fig. 7.5 Anne Frank (Creative Commons)

Fig. 7.6 Auschwitz-Birekenau women and children deemed "unfit for work" are led unknowingly to gas chamber (Creative Commons)

Fig. 7.7 The Yin and Yang in the human mind (Creative Commons)

In anthropogenic terms the Yin and the Yang pair (Fig. 7.7) is mirrored by life and death, growth and decline, good and evil. No entity can survive entirely on its own: death-seeking demons are buried deep inside the purest of hearts and love resides inside monsters. On a global scale this duality is represented by cyclic alternation between periods of renaissance and brutal wars, leading to an ultimate nuclear catastrophe.

From Space Lunacy to *Mad Max's* Fury Road

8

> *To ignore evil is to become an accomplice.*
> (Martin Luther King)

History records the self-deification as gods by kings, Pharaohs and Emperors, such as Caligula or Nero, and the open-ended ambitions of powerful rulers, recently mimicked by billionaires declaring false messianic prophecies of "intergalactic civilizations" (Fig. 8.1a) while more likely civilization and much of the living world are heading toward *Fury Road* (Fig. 8.1b).

Since the aftermath of World War II, the biggest carnage the world has ever known, large parts of affluent western societies are singing their own praises replete with Orwellian superlatives, entertained by space games featuring dark force, dead skeletons in space (Fig. 8.1a) and the Mad Max's road fury (Fig. 8.1b), overlooking the demise of the planetary life support system. With exceptions, the propagation of lies by populist and democratic agents has become paramount thanks to the power of money and claims of democracy—underpinned by "*one dollar one vote*".

A bunch of fossil fuel billionaires accompanied by a small number of hired scientists and media mouthpieces dominating the 24-hours news cycle have been denying and spreading misconceptions about the climate, promoting the spending of $billions of humanity's dwindling resources, out of the mouth of hungry children, on space adventures and future wars, cheered by media infotainment propagating hype in their chase of ratings.

From the Sputnik to the Apollo lunar landings to the exploration of Mars, religious mythologies are evolving into a space cult, alluding to a colonization of planets and a spread of human civilization in space, where presumably new terrains and possible life forms would be overwhelmed. Predictions of making life interplanetary by space enterprise proprietors such as Elon Musk, Jeff Bezos and Richard Branson, including plans for space tourism, asteroid mining and extraterrestrial human settlements would, by some estimates, be expected to cost $trillions.

© The Author(s), under exclusive license to Springer Nature Switzerland AG 2021
A. Y. Glikson, *The Fatal Species*,
https://doi.org/10.1007/978-3-030-75468-6_8

Fig. 8.1 **a** "future" of humanity—An "alternative habitat in space" (Creative Commons); **b** A *Mad Max* post-world war III world

Fig. 8.2 Stephen Hawking (Creative Commons)

All while the future of life on Earth under global warming is increasingly in question, while the price of dead planets if soaring *"A bizarre trillion-dollar asteroid worth more than our planet is now aligned with the Earth and sun."*

In 2000 Jeff Bezos, founder of Amazon and the world's richest man, has launched the reusable Blue Origin, with the aim to commence space tourism in sub-orbital flights, charging a six-figure price such as $300,000 per ticket. Further developments of returnable rockets include defense contracts with the US government and ambitions for permanent human settlement on the Moon, in partnership with NASA.

In 2002 Elon Musk, founder of Pay-Pal, developed the Space-X rocket, including 70 launches to date, signing contracts with NASA, the US Air Force and the Argentine Space Agency, including supply contract with the International Space Station. With the motto *"Making Life Interplanetary"* Space-X's ultimate goal is to send crewed flights to Mars and eventually colonize the Red Planet. *"I want to die on Mars,"* Musk has said (2017), *"just not on impact"* and *"The future is vastly more exciting and interesting if we're a space-faring civilization"* and *"You want to wake up in the morning and think the future is going to be great—and that's what being a spacefaring civilization is all about. It's about believing in the future and thinking that the future will be better than the past. And I can't think of anything more exciting than going out there and being among the stars."*

In 2004 Richard Branson launched *Virgin Galactic*, a tourist-oriented reusable 'space plane' for sub-orbital flights, having already signed some rich people on $250,000 tickets and collaborating with the UAE's sovereign wealth fund. On 13 December 2018 the VSS Unity achieved the project's first suborbital space flight, reaching an altitude of 82.7 km. In February 2019 a member of the team sat in a flight that reached an altitude of 89.9 km. According to Richard Branson's motto "*Together we open space to change the world for good*".

Such ideas have been attracting many scientists like bees to the honey. According to Stephen Hawking "*Human race is doomed if we do not colonize the Moon and Mars*", stating further: "*I am convinced that humans need to leave earth. The Earth is becoming too small for us, our physical resources are being drained at an alarming rate*" and "*We have given our planet the disastrous gift of climate change, rising temperatures, the reducing of polar ice caps, deforestation and decimation of animal species*" (Fig. 8.2).

Many have greater faith stored in science fictions, destined to conquer the solar system, even leave the galaxy once they extend their aggressive lunacy to the rest of the universe, a willful ignorance endorsing mass naivety, with faithful belief in the lies of demagogues, fact-free shamans, false prophets and mass murderers.

Space prophets appear to have little knowledge of human physiology and psychology, both intrinsically inter-connected with the terrestrial atmosphere, radiation and gravity. Should humans succeed in colonizing one or more planets, their psychology would hardly allow them to avoid destruction., as portrayed in Wells (1898) "*War of the worlds*".

It looks that just as the masters of carbon and nuclear energies are willing to compromise or give up on terrestrial species some of their colleagues suggest life can be replanted on other planets, regardless of their physics and chemistry. The lunacy of imagining forests and wildlife on Mars clash with their demise on Earth. In response to Stephen Hawking's dreams the question needs to be posed, given the human mindset how could humans avoid destroying Martian and lunar environments the way they are doing on planet Earth?

We are Earthlings, a product of terrestrial evolution; our bodies have evolved on Earth from a long lineage of species, are attuned to its gravity, atmosphere, radiation and the multitude of micro-organisms with whom we interact. The false prophecies of space colonization asserting as if alternatives exist to saving the Earth's atmosphere, oceans and biosphere, amount to irrationality.

The parallels between space mythology and religious beliefs of heaven and hell are evident. The virtuous, namely the ultra-rich and their followers, would be salvaged, while the poor, colored skinned people and those who yearn for justice would burn in hell, as Earth warms. According to Oxfam eight billionaires now own as much wealth as half the human race, spending much on space toys in ethics-free plans for planetary colonization, constituting a criminal diversion from the desperate need to save life on Earth.

It is amazing how some scientists, brilliant in their own specialties, can overlook other scientific and human disciplines. Stephen Hawking, with limited knowledge of human biology and ecology stated: "*I think the human race has no future if it doesn't go to space*", avoiding the concern humans would continue to destroy any environment they occupy. James Hansen, the pioneer of climate change science, advocates nuclear power, although if unleashed by accident or design this would render large parts of Earth uninhabitable. Scientific exploration of the planets best belongs to remote robotic micro-laboratories designed to monitor the electromagnetic wave spectrum in ways superior to individual humans.

It is a question whether a species exists in this, or any other galaxy, which has perpetrated mass extinction of species on the scale initiated by *Homo sapiens* since the mid-18th century? A critical parameter in *Drake's Equation*, which seeks to estimate the number of planets which host civilizations in the Milky Way galaxy, is L—the longevity of technological societies measured from the time radio telescopes are invented in an attempt to communicate with other planets. Estimates of L range between 10^2 and 10^7, with an estimate of $\sim 300,000$ years (Kompanichenko 2000).

According to Albert Speer, German physicists apprising Hitler of the possible construction of an atom bomb in the spring of 1942, noted a reservation by Werner Heisenberg about a potential conflagration of the atmosphere: "*Hitler was plainly not delighted with the possibility that the Earth under his rule might be transformed into a glowing star*". The same awesome possibility, fusion of atmospheric nitrogen and of oceanic hydrogen, triggering a planetary chain-reaction, has been considered a few months later by Edward Teller, Robert Oppenheimer, Arthur Compton, Hans Bethe and other physicists. New calculations indicated atmospheric conflagration was unlikely, allowing the Trinity nuclear test in the New Mexico desert to go ahead.

Planetcide emerged in the mind of Promethean pre-historic humans around camp fires, from dark recesses of the mind and their attempt to survive adversity, from Atavistic fear of death leading yearning for god-like immortality. Once the climate stabilized in the Holocene about 7000 years ago and excess food was produced, fear and aggression, its counterpart, grew out of control, constructing pyramids dedicated to immortality, celebrated by murderous orgies termed *war*, intended to conquer death in order to appease the Gods.

According to futuristic ideas, including by Yale University astronomer Gregory Laughlin, the prospect for life may be bright given continuing evolution and technological advances, allowing humans to survive in some form even long after Earth ceased to exist. Such beliefs, divorcing life from the planet, have currency among those who can afford a form of cryogenic burial.

The ideal of a glorious death in battle, the ritual sacrifice of the young as a way of gaining eternity, as expressed by an Aztec war song: (1300–1521 AD):

There is nothing like death in war,
Nothing like flowering death,
So precious to him who gives life,
Far off I see it, my heart yearns for it.

From infanticide by rival warlord baboons, to slaughtering of children on Aztec altars, to the generational sacrifice of World Wars (Fig. 8.10), the young follow leaders blindly, priests promote bigotry and crusades. Hijacking the image of Christ, a messenger of justice, they promote their self-fulfilling *Armageddon*, singing "*Hallelujah the rapture is coming*".

With virtually unlimited solid, liquid and gas carbon reserves, extractable from the Earth crust further emissions will take the atmosphere out of the ice ages back to

Fig. 8.3 Damocles sword. Damocles sits on a throne. Dionysius is standing next to him and gestures at the sword (Creative Commons)

Mesozoic-like tropical conditions, when large parts of the continents were inundated, if not to Venus-like runaway greenhouse conditions. More likely to survive are the grasses, some insect species and perhaps some bird species, descendants of the fated dinosaurs. A new evolutionary cycle may commence. Some humans may survive—where the endurance of members of the species through the extreme climate upheavals of the glacial-interglacial periods has equipped them to withstand the most challenging conditions.

The sixth mass extinction of species may be brought about, separately or in combination, by both the climate calamity and a global nuclear cataclysm. As time goes on, possibilities become probabilities and possibilities become certainties, a warming planet burdened by resource wars loses much of its life support system. Following the atomic annihilation of the inhabitants of two Japanese cities, with time the Damocles sword (Fig. 8.3) of the Mutual Assured Destruction

(MAD) strategy can only fall, where the hapless inhabitants of planet Earth are given no choice between global heating and the coup-de-grace of a nuclear winter.

Whereas the Hadron Collider experiment has been deemed to be "*safe*", pending further science fiction-like experiments yet to be dreamt by ethics-free scientists, the Earth may or may not survive. Unfortunately little doubt exists about the consequences of the continuing use of the atmosphere, the lungs of the biosphere, as an open sewer for carbon gases.

In 1986 the renowned oceanographer Wallace Broecker stated: "*The inhabitants of planet Earth are quietly conducting a gigantic experiment. We play Russian roulette with climate and no one knows what lies in the active chamber of the gun*" In the wake of this projection, as stated by Wallace-Wells (2017): "*The slowness of climate change is a fairy tale, perhaps as pernicious as the one that says it isn't happening at all, and if your anxiety about it is dominated by fears of sea-level rise, you are barely scratching the surface of what terrors are possible, even within the lifetime of a teenager today. Over the past decades, the term "Anthropocene" has climbed into the popular imagination—a name given to the geologic era we live in now, one defined by human intervention in the life of the planet. But however sanguine you might be about the proposition that we have ravaged the natural world, which we surely have, it is another thing entirely to consider the possibility that we have only provoked it, engineering first in ignorance and then in denial a climate system that will now go to war with us for many centuries, perhaps until it destroys us. In the meantime, it will remake us, transforming every aspect of the way we live-the planet no longer nurturing a dream of abundance, but a living nightmare*".

Where the Nazis constructed gas chambers for millions of victims the climate calamity is threatening to turn the planet into an open gas oven, on the strength of a *Faustian Bargain* promoted by vested interests and their mouthpieces.

From the Roman Empire to the Third Reich and its successors, the barbarism of empires surpasses that of small marauding tribes. In the name of *democracy*, *freedom* and *human rights* the powerful never cease to burn hamlets, killing the wretched of the Earth in their fields and villages, lately using land mines and drones, as in Vietnam and Afghanistan. It is among the poorest of the Earth that true charity is common, where empathy is learnt through suffering.

In the name of democracy and freedom of speech everyone, including hired mercenaries, has access to electronic networks propagating "alternative facts" contradicting the basic laws of physics and direct observations regarding the reality of global warming, the nature of pandemics, of critical life-and-death nature.

Planetcide challenges every social system, faith and ideal that humans have ever held. As in *War of the Worlds'*, individuals are crushed and cells which rebel are destroyed by the parent organism, exerting the power of *National Security*. The build-up of *too much power* (Fig. 8.4) in the name of "*defense*" culminates in *offence*.

In the story of the tower of Babel god confuses the languages of people to prevent them from reaching the sky. Planetcide is a child of Orwellian *Newspeak*, where modern societies underpinned by the military-industrial complex, merchants

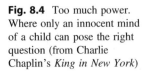

Fig. 8.4 Too much power. Where only an innocent mind of a child can pose the right question (from Charlie Chaplin's *King in New York*)

of death, subterranean drug rings and intelligence agencies, connected with financial institutions and casinos, poison their young's minds with commercial advertising and political lies. Where the aggressor portrays himself as a victim in Joseph Goebbels tradition, as he stated: "*If you tell a lie big enough and keep repeating it, people will eventually come to believe it. The lie can be maintained only for such time as the State can shield the people from the political, economic and/or military consequences of the lie. It thus becomes vitally important for the State to use all of its powers to repress dissent, for the truth is the mortal enemy of the lie, and thus by extension, the truth is the greatest enemy of the State.*". Nowadays reality is redefined by *fake news* and *manufactured consent*.

Nature is full of parasites, viruses destroying their hosts, intelligent spiders trapping their victims, sea anemones seducing their prey, members of *Homo sapiens* perfecting untruths to an art form. Defying the basic laws of physics, the peer review system and direct observations in nature, golden-hair neo-Nazi boys, hipsters and pseudoscientists lured by vested interests have become mouthpieces of air-poisoning lobbies which, for longer than forty years, continued to delay humanity's desperate efforts at mitigating the fast deteriorating state of the atmosphere.

Having lost the sense of reverence prehistoric humans possessed toward the Earth, there is no evidence civilization is about to adopt Carl Sagan's sentiment (Cosmos 1980) "*For we are the local embodiment of a cosmos grown to self-awareness. We have begun to contemplate our origins: star stuff pondering the stars: organized assemblages of ten billion-billion-billion atoms considering the evolution of atoms; tracing the long journey by which, here at least, consciousness arose. Our loyalties are to the species and the planet. We speak for the Earth. Our*

Fig. 8.5 Paradise. Jan Bruegel (Creative Commons)

obligation to survive is owed not just to ourselves but also to that Cosmos, ancient and vast, from which we spring."

Throughout the ages human endurance of fear of pain and death was in part sustained by a belief in supernatural powers the people trusted to save them, or at least ensure their survival in the afterlife. Atheist philosophy, recently manifested by *"The god delusion"* of Dawkins (2016) attacks the concept of god, points to the improbability of a supreme being, indicates how religion fuels war, foments bigotry and abuses children. Atheists, however, take little account of the deep-rooted human need for relief from pain and for guidance and support from external powers. Human dreams of peace and harmony in paradise (Fig. 8.5) are signified by the legend of the Hesperides[1] with their golden apples (Fig. 8.6).

Voltaire struggles with the concept of natural catastrophes in an otherwise ordered universe where god allows both good and the bad to coexist. Albert Camus faced an existential conflict between a belief in a God and a world of inexplicable

---

[1]The Hesperides are the nymphs of evening and golden light of sunsets, who were the "Daughters of the Evening" or "Nymphs of the West"

Fig. 8.6 The Hesperides British Museum, London. Meidias Painter, 420-400 B.C (Creative Commons)

human suffering. At the heart of the dilemma is the problem of evil. *"It seems as if, when the cause of the evil is "natural" Christians are faced with a real problem. Either God is unable to prevent large-scale unjust suffering, or God chooses not to. Moral evil is wrought by humans themselves on the world and is readily accepted as a consequence of the gift of free will"*.

Alternate spectres of Dante's Inferno, realized by torture inflicted by the inquisition, were used by the church, kings and princes to instil mortal fear, but while religious faith allowed people to endure unbearable pain, this has in turn created its own hell, as expressed by William Shakespeare *"Hell is empty and all the devils are here"* (Fig. 8.7) under the medieval inquisition (Fig. 8.8).

The adulation of *god* has been more recently substituted by the idolization of celebrities, popular actors or sportsmen, playing the "hand of god". Where the Romans were entertained they watching blood curdling killings at the colosseum, residents of suburbia satisfy their appetite watching *"whodunit"* homicide movies, cf. *Sherlock Holmes, Agatha Christie or Murder She Wrote*, devoid of humane empathy. The victims, once children loved by their parents with hope for a future, are cut out brutally.

Inherent in the *Winner Takes All* mentality is open-ended craving symbolized by the allure of gold, a token for which millions suffered and died in the gold mines and civilizations destroyed, as in the Americas and in mine disasters elsewhere.

The earth will swallow us who burrow
And, if I die there underground,
What does it matter?

Fig. 8.7 *"Hell is empty and all the devils are here"* (William Shakespeare). Panel from the Garden of Earthly Delights by Hieronymus Bosch (Creative Commons)

Who am I? Dear Lord!
All round me, every day,
1 see men stumble, fall and die.
B. W. Vilakazi

Which is one way to perish. Between 1740 and 1897 there were 230 wars and revolutions in Europe[2] (Fig. 8.9). According to Doris Lessing the most unsettling human failing is the ease by which populations can be manipulated, frightened and incited against perceived "enemies", manifested by enthusiastic marches of young recruits toward war (Figs. 8.9, 8.10, 8.11 and 8.12), many never to return. Buried deep inside the psyche are demons triggered by fear, hate and peer pressure, releasing tension. Mobs cheer mad emperors in colosseum death games coalescing into a monster of one thousand heads.

According to Taylor (2014) *"Read any book about the history of the World and it's likely that you'll be left with one overriding impression: that human beings find it impossible to live in peace with one another"*, and while justifications for the scourge of yet another bloodshed are flimsy, the generated emotions are immense.

As multitudes march cheerfully toward war (Fig. 8.10), it can be safely assumed that those who send young people to their death are rarely soiled by the mud and blood of the trenches. In 1914, German foreign minister Gottlieb von Jagow said:

---

[2]The Psychology of War: Why do humans find it so difficult to live in peace? https://www.psychologytoday.com/au/blog/out-the-darkness/201403/the-psychology-war

Fig. 8.8 Execution of Mariana de Carabajal at Mexico, from El Libro Rojo, 1870 (Creative Commons)

"Soldiers who dream of peace are an absurdity". Erich von Falkenhayn, chief of staff of the Prussian army, when he saw the cheering crowds in Berlin in August 1914, said to German Chancellor Bethmann Hollweg: *"Even if we perish, it will have been wonderful"*. The Prussian Ministry of War was full of cheering officers at the moment of mobilization; full of joy, they shouted and embraced each other. In 1914 Winston Churchill said *"If war can be avoided, I lived for nothing"*, and following Gallipoli in 1916 *"I think a curse should rest on me—because I love this war. I know it's smashing and shattering the lives of thousands every moment and yet, I can't help it, I enjoy every second of it`*.

The consequences were in the bomb-shelled craters (Fig. 8.11), not shared by the generals, and the real heroes of the war were not those marching with

Fig. 8.9 La Liberté guidant le peuple, painting by Eugène Delacroix commemorating the July Revolution of 1830 (Creative Commons)

Fig. 8.10 Soldiers cheering on the train toward the frontier in World War I

Fig. 8.11 All Quiet on the Western Front. A German soldier tries to comfort a dying French soldier. From a film based on the book by Erich Maria Remarque

Fig. 8.12 The Christmas Truce: the real heroes of World War I (Creative Commons)

enthusiasm toward the killing fields but those who dared, under the pain of death, to come out of the trenches to celebrate the Christmas Truce (Fig. 8.12).

Not too well hidden beneath the triggers for war are the vested interests of merchants of death (Engelbrecht and Hanighen 1934), consistent with the Orwellian dictum *"The conscious and intelligent manipulation of the organized habits and opinions of the masses is an important element in democratic society. Those who manipulate this unseen mechanism of society constitute an invisible government which is the true ruling power of our country"…"In almost every act of our daily lives, whether in the sphere of politics or business, in our social conduct or our ethical thinking, we are dominated by the relatively small number of persons who understand the mental processes and social patterns of the masses. It is they who pull the wires which control the public mind."* (Edward Bernays).

According to free market ideology evolved into "economic rationalism", not much exists which is not for sale—goods, services, ideas, so-called *intellectual property*, soap powder and dangerous lies—a fatal ideology for life and nature, where the the price of everything supersedes the value of nothing, extinguishing principles of justice and the ethics of the enlightenment with, so called, *economic rationalism*.

The deterioration of values is reflected by poor articulation, a language replete with empty terms and expletive superlatives hinting at an impoverishment of the intellect, including repetitive exclamations such as *"it's kind of cool, man"*, *"awesome sort of stuff"*, *"you know what I mean"*, shifting the onus of explanation on to the other party. Removed from the realities of the natural world, cerebral processes indulge in conspiracy theories inspired by streams of science fiction and *"reality TV"* shows, the origin of cognitive dissonance. Back from busy office manipulating spread sheets, shifting electrons from USB plug-ins to hard drives, couch potatoes mesmerized by fluorescent TV screens to consume infotainment propagated by commercial and political advertisers and the powers-that-be, while humans become divorced from their fast deteriorating planetary environment.

TV or not To Be

An ecstatic kid jumped high and froze
In mid-air, munching a sweet dose
Of chocolate-coated candy bar
The latest taste sensation star

Two lovers share romantic stance
Absorbed intensely in their trance
Thanks to the bank's new interest rate
Reduced by one percent to date

Smug businessmen proudly proclaim
To secretaries cheers, acclaim
The year 2000 grandest vision
Debentured Futures profit mission.

A grey haired couple now appears
Rolled over in their super-years
The man says smiling to his wife
Thank God for value-added life.

I soon switch off this box of lies
And raise my head to blackened skies
The roaring thunder, lightning flash
A world that won't be saved by cash

by Andrew Glikson

The Triumph of the Absurd

9

Think of the earth as a living organism that is being attacked by billions of bacteria
(Gore Vidal)

The smart way to keep people passive and obedient is to strictly limit the spectrum of acceptable opinion, but allow lively debate within that spectrum
(Noam Chomsky)

A majority of humans possess innocuous minds, preoccupied by hopes, beliefs and absurd imagination (Fig. 9.1), yet susceptible to misconceptions and lies, denoted by Chomsky and Hermann 1995 *as manufactured consent*. A minority, preoccupied with mercenary schemes, exert control over such innocence. Both individual and collective lives are overtaken by paradoxes, such as where ethically decent but politically naïve people's lives are overtaken by mercenary schemes or, on a larger scale, deadly conflicts ensue where technological societies clash with unwary tribal communities, as documented by Reynolds (2013).

During periods of enlightenment, such as the renaissance, humans bestowed beautiful cultural signatures on the world, yet in oblivious submission to Orwellian lies ended up overtaken by dark forces committing carnage. In such a world, portrayed by George Orwell's novel 1984, mind control and thought-crime extinguish all meaning and doubt from the language, leaving out simple dichotomies such as pleasure and pain, happiness and sadness, good and bad, ethics and evil thoughts, disabling any questioning of the dominant philosophy of the regime and reinforcing a total dominance of "*Big Brother*". History is written by the winners (George Orwell).

Unlike many organisms humans appear to have the capacity of anticipating danger well in advance, harbouring fears that may force them to toe the line through herd mentality, indoctrination, vested interests , generating irrational behaviour with criminal consequences. In the midth of battle soldiers, caught between an enemy in front and officers with drawn revolvers behind, are driven from their humanity,

A. Y. Glikson, *The Fatal Species*,
https://doi.org/10.1007/978-3-030-75468-6_9

Fig. 9.1 The absurd: Don Quixote de la Mancha and Sancho Panza, riding their mounts. An engraving based on a scene from Miguel de Cervantes 1863, by Gustave Doré (Creative Commons)

killing people they never met. In silos or submarines missile launch officers turn keys to commit the lives of hundreds of thousands humans they never knew.

For an infant growing within a bubble, detached from external reality, the world constitutes an empty reflection, nor can a child growing in front of TV shows be expected to become a healthy adult. Not much more can be hoped for from a population bombarded by social media, infotainment, fake news, alternative facts, bit coin speculators dominated by an ethics-free advertising industry and professional liars.

Lies proliferate like viruses. As expressed by Euripides, Orestes, 408 BC: "*When one with honeyed words but evil mind persuades the mob, great woes befall the state*", politics have little to do with the truth. Social and commercial propaganda screaming from *Facebook*, *Twitter*, *Netflix*, *Instagram*, *You tube*, *Snapchat*, *Wechat*, promote science fictions to become manufactured facts. A moron-dominated global disinformation jungle radiating from fluorescent screens features space robots, flag-waving androids, Hollywood heroes, Disneyland clowns, Las Vegas gamblers, vicious rapists, gun-toting brutes and Brown Shirt militias lead mass murders—all becoming everyday's virtual reality while the wretched of the Earth are strafed by drones.

According to Ernest Mayr, the renowned American biologist, intelligence is a kind of lethal mutation. Biological success, measured by the size of populations, depends on the rate of mutations. Intelligence is self-limiting lethal mutation in terms of survival of a species since, as soon as the species develops a potentially self-destructive technology, it may end up destroying itself in the process. He also adds, ominously, that the average life span of a species, of the billions that have existed, is about 100,000 years, which is almost the length of time modern humans have existed.

According to Chomsky (2010) *"What Mayer basically argued is that intelligence is a kind of lethal mutation. And he had a good argument. He pointed out that if you take a look at biological success, which is essentially measured by how many of us are there, the organisms that do quite well are those that mutate very quickly, like bacteria, or those that are stuck in a fixed ecological niche, like beetles. They do fine. And they may survive the environmental crisis. But as you go up the scale of what we call intelligence, they are less and less successful. By the time you get to mammals, there are very few of them as compared with, say, insects. By the time you get to humans, the origin of humans may be 100,000 years ago, there is a very small group. We are kind of misled now because there are a lot of humans around, but that's a matter of a few thousand years, which is meaningless from an evolutionary point of view. His argument was, you're just not going to find intelligent life elsewhere, and you probably won't find it here for very long either because it's just a lethal mutation. He also added, a little bit ominously, that the average life span of a species, of the billions that have existed, is about 100,000 years, which is roughly the length of time that modern humans have existed."*

Since the second half of the twentieth century humans have entered the electronic age where a majority of technological developments are taking place. Megalomaniac billionaires and their armies of lawyers undermine what is left of the sixteenth century enlightenment, reinventing medieval evidence-free conspiracy theories propagated by moronic channels as alternative facts akin to the dangerous Q-Anon type cults.

Where power supersedes truth freedom is translated into a nightmare no amount of empty superlatives can conceal. In his novel *Brave New World* Aldous Huxley (1932) portrays a dystopian infotainment-controlled society bearing essential similarities to that nowadays propagated on a myriad electronic devices inhabited by mouthpieces, talking heads and inane celebs, oozing showy charm and frivolous falsehoods devoid of a moral compass. As the gap between commercial advertising and dangerous propaganda is blurred, suburbia international is streamlined to perpetrate the next bloodbath, killing coloured people in remote parts of the world, supposedly in the name of freedom and democracy.

Much like prehistoric humans perched around campfire flames, multitudes are captivated by fluorescent screens, subject to the propagation of mixtures of fact and fiction, truths and dangerous untruths poisoning people's minds toward homicide and genocide. Contradictory forces are in play, including the rags to riches myth where everyone can be a millionaire. The advent of electronic warfare systems, reinforced by missile and drone fleets, is leading to an Orwellian-like 1984 global mind control system (Fig. 9.2), ensuring the suppression of human *"thought-crimes"*.

Psychological preparation of societies for war is conducted through the vilification of potential "enemies". A new cycle of fascism and a potential world war is emerging late in the twentieth century and early in the twenty-first century, at the

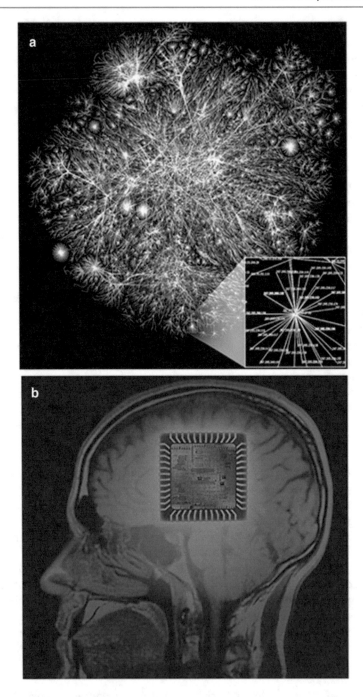

Fig. 9.2 Global mind controls. **a** The connections and pathways of the internet emulate pathways of neurons and synapses in a global brain; **b** A computer brain (Creative Commons)

very time when the world needs to cooperate to try and save its planetary life support systems from a climate catastrophe as well as urgently dismantle its arsenal of nuclear weapons. Fascism, emerging from dark recesses of the human mind, criminal sociopathy and survival of the fittest fundamentalism (social Darwinism), coalesces in violent movements promoting death in all its forms.

Chaos reigns, while Homo "*sapiens*" continues with the suicidal transfer of every accessible molecule of carbon from the Earth crust to the atmosphere, the auto-da-fe of the living Earth (Fig. 8.7) proceeds as the warming oceans are rising, flooding the cradles of civilization of the great river valleys—the Nile, Mesopotamia, the Hindus, Ganges, Mekong and Yellow River—history is nearing its nadir.

Sporting 4-wheel SUV[1]s and mobile *Trailblazers* allowing virtual escape, waving national flags, people disperse like ants before the rain, lost in a virtual world of *Faceless Books and on-line Tweets*. They jet around the world to frequent mass circuses where *World-Number-One legends and celebrity Icons* write themselves into history. A Commercial screams: *See the world before it is too late.*

Jo Star, an aerobatics ace pilot of right stuff fame prides himself in emitting thousands of tons of carbon dioxide each year, burning the liquid deposits of Jurassic plankton and microflora that populated the Tethys Sea. Self-righteous billionaires play with rockets, satellites and drones; relic semblances of democracy are hijacked by fossil fuel tycoons; snake oil merchants drug the young; human ascendancy supremacists perfect deception as a form of art; few thousand of the mega-rich amass greater wealth than 4.6 billion people in national systems referred to as democracies.

By the second decade of the 21st century the rates of greenhouse gas rise supersedes any since 66 million years ago, tracking above the stability limit of the Antarctic ice sheet. The base of the marine food chain and algal reefs are stressed by warming acid water. The rising oceans threaten coastal plains, low river valleys and urban hubs of civilization.

The writing has been on the wall ever since *Homo erectus* mastered fire, perhaps about 1.7–2.0 million years ago. Political and media heads debate global warming: A fat senator declares: "*To be quite honest it's a challenge, you know, we just need to grow the economy, going forward, you know, it is the twenty-first century, you know*". A 50,000-strong crowd at the MCG[2] cheer, most of them oblivious to what their troops are doing, killing poor farmers in faraway lands, the demise of their home planet, and the fate of the species. How many of them care?

Primordial howls Shriek, what passes for music betrays an unconscious knowledge of what's to come. Pompous hubris reverberates "*Global warming is from the sun; the Earth is cooling; carbon dioxide is plant food; the greenhouse effect is a United Nations conspiracy; clean coal; coal is good for the poor of the world; drill baby drill; it's the economy stupid, the market force decides; corrupt scientists seek research grants*".

---

[1]SUV—Special utility vehicle.
[2]MCG—Melbourne Cricket Ground.

Executives clad in designer labeled suits raise a toast to corporate profits; media channels exude canned laughter; school children line up in parliament's Marble Hall, tiny tots with heads looking up at portraits of leaders supposed to protect their future; downtown a youth tangled in i-phones and i-pads twits a memo; a pregnant woman pushing a pram is passing by; to the sound of heavy metal music a girl exclaims: "*It is awesome, you know, it's cool, you know, it's groovy, you know*".

"No joy could sate him, and suffice no bliss, to catch but
shifting shapes was his endeavour, the lastest, the poorest,
emptiest moment - this, he wished to hold it fast forever."
(Mephisto, in Goethe's *Faust*).

The Executive

He flies golden wings class
Wearing a smart pinstriped suit
A designer label silk tie, buttons brass
By monsieur Pier Cardin, he lies
While sipping Martini on ice, a caress
With the smile of a charming hostess

He stays in a luxury inn
On soft-leathered VIP lounge
Pressing on button push gadgets
To meet his whims and demands
He Rings a credit card-paid guest
To satisfy his bodily lust

He summons directors, a weird bunch
In a skyscraper's bar for a lunch
Finalising contracts, shady deals
That the lives of pure mortals will seal
A few blocks downtown in the traps
An old man is scavenging scraps

He is driven in a black limousine
A silent chauffeur wears a grin
A door open, a colourful page
Genuflect, as if to a lord or a sage
He returns home to a wife's secret hate
And the resentment of two kids

One child an addict in need
The other a dreamer of anarchy deed
He kisses them all on the cheek
Sets in front of TV, not too meek
But in his hand he keeps clutching
Pandora's black box secrets

by Andrew Glikson

It becomes obvious no one is telling the truth and that the language, intimately intertwined with thought, is regressing. People, compromised and co-opted, pay lip service to the crisis, too afraid to resist the insanity, except for the children led by Greta Thunberg.

Memories fade with every passing generation, forgetting World War II where some 70–85 million people were killed, and once again, fascism is on the rise around the world. Firestorms engulfed the forests of southeastern Australia, California, the Amazon, even large parts of the Arctic. Millions of animals perish. 24 hour news cycles wipe out memories.

Tribal aggression inherent in ancestor primates, once equipped with carbon-emitting and nuclear devices, is leading the species to extinction, taking many other life forms with it. With some exceptions human institutions appear to be adverse to scientists' warnings. Individuals endangering the status quo and group-think are ignored or penalized. Anti-nuclear and climate scientists are labelled "alarmists", lose positions in the bureaucracy and even in academia. Nor do scientists always have the courage to report the full consequences of global heating tracking toward 2 °C on the continents and projected to near 4 °C later in the century.

The realization that life on Earth is in danger is more than many people are prepared or able to contemplate. For many an inherent optimism, fear or belief-based bias turn out more powerful than evidence-based reason. As stated by Albert Einstein's dictum *"we cannot solve our problems with the same thinking we used when we created them."*

According to Bowden (2013) civilization and war were born around the same time, challenging the belief that the more civilized the world becomes the less likely it is the resort to war to resolve disputes. During brief intervals between periods of global devastation, such as during the Weimar Republic between World Wars I and II, societies temporarily return to their senses and a semblance of the enlightenment temporarily resumed. A similar situation pertains in much of the world since World War II, apart from genocidal atrocities in several parts—Korea, Viet Nam, Iraq, Afghanistan, persisting to the early twenty-first century. The enlightenment, born in Europe in the 17-18th centuries, concomitant with the rise of the sciences, reversed during world wars I and II, part recovering following WWII, is fast receding as the nuclear race and global heating threaten the very survival of human civilization and of much of the biosphere.

According to James Waterman *"When fascism comes to America it will be wrapped in the flag and carrying a cross"*. In his 1935 novel *"It Cannot Happen Here"* Sinclair Lewis (1935) pits liberal complacency against popular fascism. It shows that yes, it really can happen here. In an article titled *"American Fascism: It Has Happened Here"* Sarah Churchwell writes: *"Nostalgia for a purer, mythic, often rural past; cults of tradition and cultural regeneration; paramilitary groups; the delegitimizing of political opponents and demonization of critics; the universalizing of some groups as authentically national, while dehumanizing all other groups; hostility to intellectualism and attacks on a free press; anti-modernism; fetishized patriarchal masculinity; and a distressed sense of victimhood and collective grievance. Fascist mythologies often incorporate a notion of cleansing, an*

Fig. 9.3 The Yin and Yang

Fig. 9.4 Darth Vader—the
face of fascism (Creative
Commons)

*exclusionary defense against racial or cultural contamination, and related
eugenicist preferences for certain 'bloodlines' over others."*

The killer gene is never hidden far below the surface, looking for a flag to
express itself, which racist movements provide, where the worst human instincts
find a cause and control is maximized in the hands of dark forces, where death
assaults the life force, symbolized by the Yin and Yang (Fig. 9.3).

Two swastika-adorned leather-clad bike riders are passing by (Fig. 9.4):

Whining Bikes black smoke exude
Swastika-clad black death elude
Sun glasses-mounted bald head
Drooped moustache, a shadow beard
Black leather silver chains of fear
Heart tattoos a dagger bear
Steel spike braced boots, flares

Scream of toughness, hate declares
Yet no killer's gun can weave
The web of life that seeds conceive

by Andrew Glikson

Burning the Lungs of Earth

10

The paleoclimate record shouts to us that, far from being self-stabilizing, the Earth's climate system is an ornery beast which over-reacts even to small nudges.
Wallace Smith Broecker

We're simply talking about the very life support system of this planet.
Hans Joachim Schellnhuber (2009)

Earth's rate of global warming Is 400,000 Hiroshima bombs a day 365 days per year.

James Hansen (2016)

The consequences of the criminal blindness of governments to warnings issued by climate science over 30 years, specifically in James Hansen's testimony to the US Congress, are upon us. The year 2020 was <u>one of the three warmest</u> in the climate record, experiencing massive wildfires in Australia, Brazil and the west coast of North America, drought in South America, flooding in Africa, China and Southeast Asia and a record hurricane season in the Atlantic. By 2050, the living space of more than a billion people may be threatened. Serious consequences ensue for the fauna, including polar bears which starve due to the shrinking Arctic ice sheets, krill living under ice and suffering from shrinking sea sheets, corals extinguished by bleaching, acidification, marine mammals which depend on the krill, forest fires decimating mammal, lizards and bird habitats.

The history of Earth includes at least five major mass extinctions (Stanley 1987) defining the termination of eras, including the Ordovician, Devonian, Permian, Jurassic and Cretaceous. Each of these events coincided with either extra-terrestrial impacts, or massive volcanic eruptions, or massive release of methane, ocean unoxia, involving the release of greenhouse gases and inducing changes in atmospheric composition and temperature (Figs. 10.1 and 10.2). With the exception of

Fig. 10.1 The change in state of the planetary climate since the onset of the industrial age in the eighteenth century

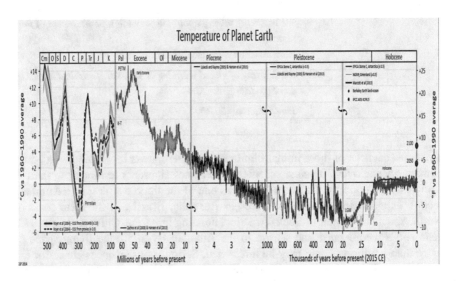

Fig. 10.2 Temperature trends for the past 550 Million years and potential geo-historical analogs for future climates (Creative Commons)

the relations between methanogenic bacteria and methane eruptions, the Sixth mass extinction of species is a novelty in planetary history, as the biosphere is disrupted by a living organism, a carbon-emitting biped mammal.

Three million years ago upper Pliocene temperatures and sea levels were higher than the early Pleistocene by 2–3 °C and about 25 m, respectively (Burke et al. 2018). The accentuation of climate oscillations was followed by the appearance of Homo erectus. The stabilization of the climate between about 10,000 and 7,000 years ago saw the Neolithic and agricultural civilization take hold. Anthropogenic processes during this period, denoted as the Anthropocoene (Steffen et al. 2007), have led to deforestation and the demise of species, ever increasing carbon pollution and temperature rise (Figs. 10.1, 10.2 and 10.3), acidification, radioactive contamination and a growing threat to the Earth's life support system (Glikson 2013).

Over geological time, through photosynthesis, the atmosphere has developed an oxygen-rich carbon-constrained composition, acting as the lungs of the biosphere, allowing emergence of oxygen-breathing animals. Planetcide results from the anthropogenic release since the industrial revolution about ~1750 AD of more than 375 GtC (billion tons carbon), emitted through the extraction and combustion of the buried products of ancient biospheres, threatening to return Earth to conditions which preceded the emergence of large mammals in the early Eocene.

The sharp glacial-interglacial oscillations of the Pleistocene (2.6 million to 10,000 years ago), saw rapid mean global temperature changes of up to 5 °C over a few millennia. This included abrupt transient stadials cooling events over a few years (Steffensen et al. 2008), which required humans to develop an extreme adaptability. Not least, of all the life forms, *Homo sapiens* succeeded in mastering fire, proceeding to manipulate the electromagnetic spectrum, split the atom and travel to other planets, a cultural evolution overtaking biological evolution.

The absurd capacity developed by humans to create and destroy at the same time is culminating with the demise of the environment that allowed them to flourish. Possessed by a conscious fear of death and craving a god-like immortality, there is no murderous obscenity some are not willing to perform, on their own initiative or under coercion. Where short-term mercenary acts overtake ethical scruples, such as the mining of air poisoning coal or manufacture of nuclear triggers, the root factors

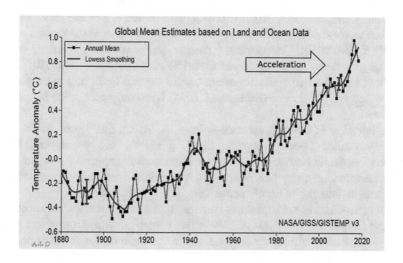

Fig. 10.3 NASA Global mean temperature estimates for Land and ocean 1880 to present, with base period 1951–1980. The solid black line is the global annual mean and the solid red line is the smoothed five-year (NASA)

for the transformation of tribal warriors into button-pushing automatons capable of destroying the world around them remain inexplicable.

Inherent in the enigma are little-understood top-to-base processes that govern the behaviour of populations, including actions contrary to the imperative of survival. Parallels in the natural world include self-destruction by Carpenter ants and termites (Autothysis). The rules of social behaviour, conforming, dissenting, constructive or destructive are obscure. According to George Ellis (2012) *"although the laws of physics explain much of the world around us, we still do not have a realistic description of causality in truly complex hierarchical structures"* (Ellis 2002).

The life support systems of the biosphere are threatened by the inexorable rise in the level of greenhouse gases and consequent rise in temperature at the Earth surface, by an average of more than 1.14 °C since 1880, currently tracking toward 2 °C. These values take little account of the masking effects of the transient mitigating effects of sulphate aerosols in the range regionally from -0.3 to -1.8 Wm^{-2}, which pushes actual local temperatures above 1.5 °C. Following the current acceleration (Fig. 10.3) mean global temperatures are likely to reach 2 °C by 2030, 3 °C by the 2050s and 4 °C by 2100, leading to a prevalence of heat waves and fires.

Warming of large ocean regions to $\sim$ 700 m deep levels reduces the ocean's ability to absorb CO_2, leaving more CO_2 in the atmosphere, with consequent accelerated warming. As ocean temperatures rise oxygen is expelled, leading to production of methane and hydrogen sulphide, poisonous for marine life and, once released to the atmosphere, terrestrial life as well.

Models projecting global warming as a linear to curved trajectory, as outlined by the International panel of Climate Change (IPCC), take little account of amplifying feedbacks, incipient stadial cooling effects and the shift of climate zone boundaries. As temperatures rise in the Arctic and the circum-Arctic jet stream boundary is increasingly weakened (Fig. 10.4a) allowing penetration of freezing fronts from the north and warm air masses from the south, reported in *"fires in the Tundra and the Arctic"* (NASA 2019). Concomitantly climate zones, such as the Sahara tropical zone, migrate away from the equator and toward the poles (Fig. 10.4b). The migrations shift the Earth climate pattern toward tropical Earth state such as during the Miocene and the Eocene (Burke et al. 2018) when only smaller ice sheets or no ice existed.

The injection of freezing air masses from the Arctic into North America and Europe, so-called "Beast from the East", producing heavy snowfall, further weakens the Arctic boundary. Cooling of large surface areas of the ocean by ice melt water flowing from Greenland and Antarctica, overlying warmer water in depth lead to irregular to erratic ocean temperature imbalance.

The concentration of greenhouse gases in the atmosphere, now exceeding 500 ppm CO_2-equivalent (NOAA 2019) when combined with methane and nitrous oxide, is generating amplifying feedbacks, releases more carbon to the atmosphere from warming oceans, methane leaks, desiccated vegetation and fires. The gulf between the amount of carbon which keeps accumulating and which can be down-drawn from the atmosphere has been underestimated. Once the greenhouse

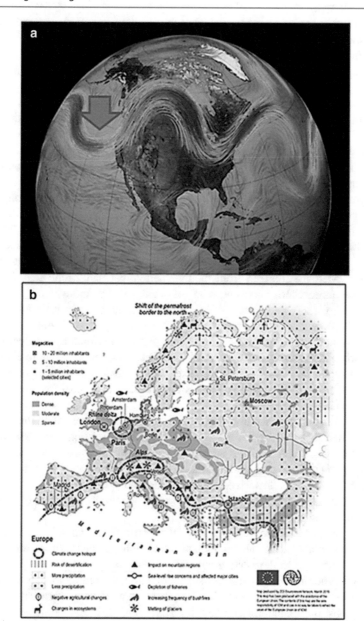

Fig. 10.4 a Increased undulation of the Arctic boundary zone, allowing penetration of cold air masses southward (NASA); **b** the migration of the Sahara arid climate zone northward into southern Europe. Note the drying up of Spain, Italy, Greece and Turkey and the increased precipitation in Northern Europe (by permission)

and radioactive genies have escaped from the Pandora box, there is no alternative but to adopt cooperation and abandon competition for the sake of survival, had it not been for the industry of lies masking reality. Technological advances such as carbon capture and storage may provide stopgap pause if not a change in human predator behaviour.

The Earth's atmosphere contains about 850 gigaton of carbon (GtC), with about 375 GtC emitted since 1750, approximately 50% of which remain in the atmosphere. Compare with 1400 GtC frozen in permafrost, mostly in the Arctic and sub-Arctic, where more than 100 GtC may be released by melting. Should a large part of the existing permafrost thaw, the Earth could experience dramatic, fast and dangerous warming (Glikson 2018).

The Arctic contains vast amounts of carbon in the form of methane accumulated during the Pleistocene ice ages, with a greenhouse trapping up to 80 times more heat in the atmosphere than carbon dioxide within a 5 year period and 72 times more within a 20 year period. Methane (CH_4) is the gas considered responsible for some of the largest mass extinctions in the history of Earth at 251 and 56 million years ago. Methane released from melting permafrost and Arctic sediments has raised the atmospheric concentration of methane by more than three-fold, from <600 to 1800 parts per billion. When emitted the radiative intensity of methane is more than ~ 80 times that of CO_2, declining to ~ 25 times over time. The release to the atmosphere of a significant part of the stored carbon (permafrost 900 billion ton carbon [GtC]), peatland 500 GtC, vegetation prone to fires (650 GtC), as well as the ever growing gas leaked and emitted from fracking (Fig. 10.5), is sufficient to shift most of the Earth's climate into a tropical to hyper-tropical fire-prone state.

As the planet warms, wildfires become more frequent, accelerating the warming process. The 2019–2020 wildfires in Australia have unleashed about 900 million tons of carbon dioxide into the atmosphere, the equivalent to nearly double the country's annual fossil fuel emissions. Sea level rise is projected to flood the very cradles of civilization, low river valleys, delta and coastal plains, vital to food production, estimated to displace 187 million people initially and more over time as major coastal cities are flooded (Fig. 10.6).

Rising atmospheric and ocean energy levels in tropical and subtropical regions of the Earth, notably island chains such as the Caribbean, the Philippines and south Pacific islands, lead to devastating cyclones, typhoons and hurricanes (Figs. 10.7) wreaking havoc on former paradise islands of Earth. Rising temperatures in tropical, subtropical and intermediate Mediterranean climate zones threaten to render large regions unsuitable for agriculture. High humidity and temperatures may exceed human physiological tolerance and where wet-bulb temperature reaches 35 °C marks the peak human tolerance, rendering areas uninhabitable.

The shift of climate zones (Fig. 10.4b) can only render large parts of Earth uninhabitable. The Earth's fast diminishing resources, urgently needed to feed hungry children, are being squandered on wasteful nonsense such as the futile attempts at colonizing other planets as a source of entertainment for the wealthy.

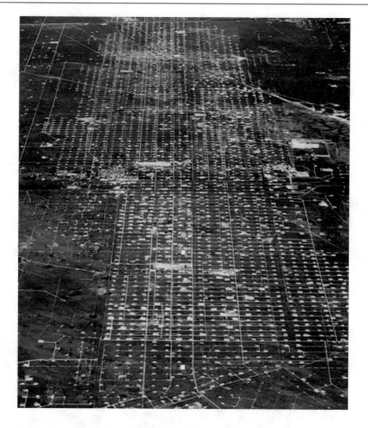

Fig. 10.5 Oil and gas fracking sites in Wickett, Texas. (Creative Commons)

A cover-up or minimization of the climate crisis by mouthpieces of vested interests, political operators, compliant mainstream journalists, anti-science ideologues (Oreskes and Conway 2015) and a small number of corrupt scientists possessing loud megaphones, has blurred the message in the eyes of the public. The complexity of the climate system, including Milankovic glacial-interglacial cycles, solar sun-spot cycles, short term variations such as the El Nino-La Nina cycles and extreme weather events (Fig. 10.8), may be confused with longer term trends. The migration of climate zones and the freezing fronts emanating from the Arctic (the "beast from the east" phenomenon) are superposed on and may obscure these cycles.

Studies of the history of the atmosphere–ocean system indicate an unprecedented rate at which atmospheric CO_2 levels are rising, i.e. by more than ~ 2 ppm/year, reaching 412 ppm by 2020, the highest level since the Pliocene 2.6 million years ago when the Arctic was ice free (Ogburn 2013). The accelerated rate of glacial melt in Greenland and West Antarctica may lead to little remaining ice toward the

Fig. 10.6 Flooded statue of Liberty By Permission, Getty Images

Fig. 10.7 Hurricane Florence churns towards the US East Coast (NOAA)

end of the century, with consequent sea level rise on the scale of many meters and catastrophic consequences for coastal and river valleys population centres.

In private conversations many climate scientists express far greater concern than they do in public (Glikson 2016) at the scale and pace of global warming and its consequences. A study titled '*When the End of Human Civilization Is Your Day*

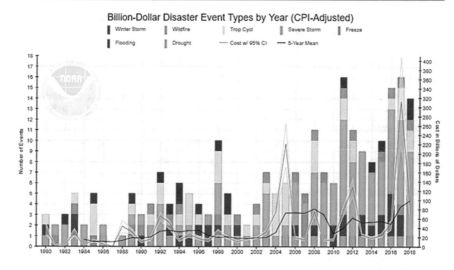

Fig. 10.8 The number, type and annual cost of U.S. billion-dollar disasters from 1980–2018. NOAA Climate.gov

Job' cites, "*Climate scientists have been so distracted and intimidated by the relentless campaign against them that they tend to avoid any statements that might get them labelled 'alarmists', retreating into a world of charts and data*". By analogy to medical scientists advising a patient of a cancer prognosis, with possible remission should the patient refrain from smoking, some patients would plunge into depression while some people, aided and abetted by the tobacco industry, will criticise and attack the doctors. It is not uncommon to hear people criticising climate scientists for not telling them more about the climate, although when they are told, they recoil. There is a heavy price to be paid by those who, displeasing the authorities, keep alerting the public. For even though the warning is based on hard evidence, scientists are denounced as "alarmists", "warmists" and "scaremongers", including accusations of "pagan emptiness" by church leaders.

Some climate scientists tend to regard the IPCC-based climate consensus as too optimistic but mostly they tend to be shunned by the media, as conveyed by Noam Chomsky: "*It's interesting that these (public climate) debates leave out almost entirely a third part of the debate, namely, a very substantial number of scientists, competent scientists, who think that the scientific consensus is much too optimistic. A group of scientists at MIT came out with a report about a year ago describing what they called the most comprehensive modelling of the climate that had ever been done. Their conclusion, which was unreported in public media as far as I know, was that the major scientific consensus of the international commission is just way off, it's much too optimistic ... their own conclusion was that unless we terminate use of fossil fuels almost immediately, it's finished. We'll never be able to overcome the consequences. That's not part of the debate*".

Climate scientists have lost employment in government and even academic institutions, have been abused, threatened and potentially face McCarthy-type witch hunts by those who deny the science. Often scientists are shunned by family and friends fatigued of gloom and doom. In a logical twist those who deny climate change refer to climate scientists as "*warmists*".

Little encouragement can be gained from the non-binding promises such as those emerging from the Paris conference, which James Hansen described as a "fraud". The unbearable knowledge that global warming to 3 and 4 °C will spell the demise of numerous species and of civilization is casting dark shadow on daily life.

The all too human characteristics, self-righteousness and double standard, prevent humanity from reaching meaningful binding agreements to dismantle the nuclear chimera and attempt to reverse the climate calamity.

Who or what will defend the Earth?

I Am Shiva

11

I have no curiosity as I can imagine the worst.
(Gore Vidal).

Life on Earth is shielded from galactic and cosmic radiation, solar energetic particles and ultraviolet and x-ray, whose enhancement or the removal of protective atmospheric and ionospheric shields, can seriously harm organisms. As he witnessed the first detonation of a nuclear weapon on July 16, 1945, a line of Hindu scripture ran through the mind of Robert Oppenheimer: "*Now I am become Death, the destroyer of worlds*".

At 08:15 am on the 6 August 1945 a second sun rose over the city of Hiroshima (Fig. 11.1). As reported by Wilfred Burchett (2015) "*I write this as a warning to the world, the city is reduced to reddish rubble and people dying from an unknown atomic plague*". More than 20 years later John Pilger (2020) wrote: "*the shadow on the steps was still there. It was an almost perfect impression of a human being at ease: legs splayed, back bent, one hand by her side as she sat waiting for a bank to open. At a quarter past eight on the morning of August 6, 1945, she and her silhouette were burned into the granite. I stared at the shadow for an hour or more then I walked down to the river where the survivors still lived in shanties. I met a man called Yukio, whose chest was etched with the pattern of the shirt he was wearing when the atomic bomb was dropped. He described a huge flash over the city, "a bluish light, something like an electrical shock", after which wind blew like a tornado and black rain fell. "I was thrown on the ground and noticed only the stalks of my flowers were left. Everything was still and quiet, and when I got up, there were people naked, not saying anything. Some of them had no skin or hair. I was certain I was dead.*"

As if this was not horrible enough, between 1946 and 1961 the US tested 1032, the Soviet Union 727, the UK 88 and France 217 nuclear devices, with a total yield equivalent of 517 megaton TNT. The biggest one, the 50 megatons TNT "*Tsar Bomba*" detonated in the Arctic Novaya Zemlya islands. Between 1946 and 1958, the United States conducted 67 nuclear tests in the Marshall Islands, the equivalent

© The Author(s), under exclusive license to Springer Nature Switzerland AG 2021
A. Y. Glikson, *The Fatal Species*,
https://doi.org/10.1007/978-3-030-75468-6_11

Fig. 11.1 Robert Oppenheimer, right, with Albert Einstein in 1947 (Creative Commons)

Fig. 11.2 Shattered world. **a** Guernica by Picasso; **b** bombed Dresden 1945; **c** a hibakusha of Hiroshima; **d** the Castle Bravo hydrogen bomb test, Bikini Atoll, 1956

Fig. 11.3 a The President and the nuclear war codes contained in a bag handled by an assistant; **b** The Castle Bravo hydrogen bomb test (Creative Commons)

of more than one Hiroshima every day for 12 years. 23 of these tests were at Bikini Atoll, and 44 near Eniwetok Atoll, with fallout spread throughout the Marshall Islands, with catastrophic human and environmental consequences. The Bikini Atoll remains poisoned earth. There were no birds.

With the power of the sun in their hands, it could be hoped human attitudes would change. Tragically, as conveyed by Albert Einstein: "*the splitting of the atom has changed everything, save our modes of thinking, and we thus drift toward unparalleled catastrophe*" (Figs. 11.2, 11.3).

The Swan Song

12

> Humans live in realms of perceptions, dreams, myths and legends, in denial of critical facts, waking up for a brief moment to witness a world as beautiful as it is cruel.
> (Eli)

In antiquity people believed the swan, at the end of its life, was allowed to sing a most beautiful song just before it dies (Fig. 12.1). Prehistoric humans, a species which has discovered fire, learnt to paint images on cave walls, emerging from the last glacial cycle with great artistic achievements betraying spiritual aspirations evolved for longer than a million years around camp fires, watching the flames, yearning for immortality (*Climate, Fire and Human Evolution; Glikson and Groves, 2016*).

Life emerged from the interaction between carbon, oxygen, hydrogen combined in rocks, water and air (Fig. 12.2) in transient reversal of entropy, governed by yet little-deciphered laws, progressing from volcanic accretions, enriching the oceans and atmosphere with minerals and gaseous products, allowing the onset of photosynthesis (Fig. 12.3) and culminating with the advent of life as intelligent swarms (Figs. 12.4 and 12.5).

Inherent in ancient scripts is a human claim for supremacy over nature as "*chosen people*", with "god on their side", as expressed in *Genesis "And now we will make human beings; they will be like us and resemble us. He created them male and female, blessed them, and said, Have many children, so that your descendants will live all over the earth and bring it under their control"* … "*They will have power over the fish, the birds, and all animals, domestic and wild large and small, so <u>god</u> created human beings, making them to be like him*". For millennia women create and nurture life, while men prepare for the next carnage, decipher the laws of physics, emit carbon and split the atom in preparation for the demise of the Earth's life support systems.

© The Author(s), under exclusive license to Springer Nature Switzerland AG 2021
A. Y. Glikson, *The Fatal Species*,
https://doi.org/10.1007/978-3-030-75468-6_12

Fig. 12.1 Black swan and
cygnets (Creative Commons)

With God on our side

(Bob Dylan)

But now we got weapons

Of chemical dust

If fire them we're forced to

Then fire them we must

One push of the button

And a shot the world wide

And you never ask questions

When God's on your side

So now as I'm leaving

I am weary as Hell

The confusion I am feeling

Aren't no tongue can tell

Fig. 12.2 Fire and water. Lava flow from Mt Kilauea, Hawaii, toward the ocean (Creative Commons)

Fig. 12.3 Stromatolite, 3.49 billion years old, Dresser Formation, Pilbara Craton, Western Australia Photograph by Andrew Glikson

Fig. 12.4 Worker bees with Queen (Creative Commons)

The words fill my head
And they fall to the floor
If God's on our side
He'll stop the next war

For longer than one million years, gathered around campfires during the long nights, mesmerized by the flickering life-like dance of the flames, prehistoric humans acquired imagination, premonitions of death, cravings for immortality, yearning for omnipotence, creating supernatural spirits in their minds. It is not known whether animals develop concepts of the future and suffer fear, except at the moment of danger. But humans recognize death even from the distance and this can obsess their mind, leading to a conflict between fear and reason, between the intuitive reptilian brain and the neocortex. As stated by Sherefkin (2016) *"To be mortal is the most basic human experience and yet man has never been able to accept it, grasp it, and behave accordingly. Man doesn't know how to be mortal"*. Humans wake up for a brief moment, in denial of critical facts, to witness a world that is as beautiful as it is brutal, applying their minds to the realm of perceptions, dreams, myths and legends.

Faith in supernatural powers allows humans to bear anguish, terror, misery, hunger and pain, but hardly to come to terms with death. Holding a mortally injured hunter in their arms, the reality of death is dawning on his comrades, with visions of the afterlife emerging in their grief. Hominids try to overcome the fear of death through victory, killing animals and humans in sacrificial tribal orgies, performing burials, practising cannibalism and launching wars. Individual humans may exude

Fig. 12.5 Cathedral termite mound. Litchfield National Park, Northern Territory, Australia (Creative Commons)

innocence and beauty, in triumph over subconscious demons, but as collectives may become prey to dark forces—the monster of a thousand heads.

Human achievements at climbing mountains ("because it is there", George Mallory), crossing the oceans, exploring the poles and reaching the terrestrial planets aim at conquest as a self-fulfilling ideology, but individual lives can be short and often unkind, a passing mirage where familiar faces and memories appear and disappear like shadows.

Criticism of the notion of human superiority in a hierarchy of nature (Jensen 2016), vis-à-vis the huge tapestry of nonhuman life, highlights the intelligence and sentience of nonhuman life. Environmental movements are often limited to the way ecological changes affect humans and their economy, with little or no regard for other creatures, but once we separate ourselves from the rest of nature we end up in positions <u>against</u> nature.

Inherent in modern philosophy, in particular in western cultures, is the assumption of *free will*, freedom where individuals can make choices in their own right. On the other hand, individual freedom as symbolized in western movies by the lone cowboy riding into the sun, contrasts with the coordinated swarm mentality in superorganisms controlled by central minds, whether the queen ant or an Adolf Hitler. While individual members of the swarm may possess unique attributes, the swarm is dominated by collective behaviour patterns, allowing it to supersede the mental qualities of individual members of the swarm.

Prevalent among human populations is a herd mentality based on fear or admiration of authority, where tribal clans are ruled by powerful fighters and nations by arms and money. Where individuals expressing doubts or rebelling are expelled or destroyed. In the modern world lawless conmen, criminal cabals and the mafia, acquiring wealth and power can penetrate the highest level of government, perpetrating corruption. Such horrid behavior takes advantage of the naïve submission of populations to propaganda and coercion. Notable exceptions have been manifested through revolutions, which not uncommonly destroy their own members, and by rebels such as Spartacus (Fig. 12.6a) and Che Guevara (Fig. 12.6b). In the 20th century heroes in the fight for justice like Martin Luther King and Ruth Bader Ginsburg (Fig. 12.7), scientists like David Attenborough, James Hansen, Helen Caldicott and Frank Fenner (Fig. 12.8) redeem the blindness of and lack of courage of so many around them.

According to Diane Halpern's book *"Sex Differences in Cognitive Abilities"* functional relationships exist between the biological functions and mental processes of the sexes. These include distinctions in the emotional and cognitive motivations such as the strong male tendency toward violence, consistent with their need for hunting prowess, while the females tend towards empathy, consistent with their role as nurturers.

In the futuristic book *Last and First Man* by Olaf Stapledon, members of an advanced human species mourn the future of Earth, trying to disseminate human seeds into space. By contrast, when confronted with the fatal consequences of carbon emission, the species *"sapiens"* continues to transfer every extractable molecule of carbon from the Earth crust into the atmosphere, the lungs of the biosphere (Fig. 12.9). Ensuring the demise of the planetary biosphere, the species has reached its moment of truth, a paradox symbolized by a candle which, while lighting the dark, is burning itself to extinction (Fig. 12.10).

Periods of murderous conflicts, crusades and imperial wars are followed by brief progressive intervals, such as during the Ptolomean era in Egypt, the Pericles rule in Athens, the Renaissance or the Weimar Republic. Periods of deep humanity and great artistic beauty emerge in-between wars, as in the artistic works of geniuses like Charlie Chaplin (Fig. 12.10) and Bob Dylan (Fig. 12.11).

With civilization and war intertwined, it seems enduring peace is only possible among small remote communities, where people acquainted with toil and sufferings learn to share and to extend generosity to strangers, as in Peter Seeger's song Guantanamera.

Fig. 12.6 Rebels. **a** Spartacus—A mosaic depicting a gladiatorial fight. From the House of the Gladiators, Kourion, Cyprus; **b** Che Guevara (Public Domain)

Martin Luther King   Ruth Bader Ginsburg

Fig. 12.7 Heroes of justice. **a** Martin Luther King; **b** Ruth Bader Ginsburg

Guantanamera

The words mean, I am a truthful man

From the land of the palm trees

And before dying, I want to share the poems of my soul

My poems are soft green,

My poems are also flaming crimson

My poems are like a wounded fawn

Seeking refuge in the forest

The last verse says

"con los pobres de la terra

With the poor people of this earth

I want to share my fate

The streams of the mountain

Pleases me more than the sea"

Fig. 12.8 Heroes of science; **a** David Attenborough; **b** James Hansen; **c** Helen Caldicott; **d** Frank Fenner

Humans wake up for a brief moment from an infinite universal slumber to witness a world which is as cruel as it is beautiful, a biosphere dominated by the food chain. They live in a realm of perceptions, dreams, myths and legends, in denial of critical facts (Koestler 1978). Men compete for food and women, who are not uncommonly raped and murdered, suffer and die in child birth. A few moments of joy and happiness brighten their lives. If looking into the sun may result in blindness so, metaphorically or according to yet little-understood laws, the deep insights into nature as reached by "*sapiens*" may bear a terrible price.

Existential philosophy allows a perspective into ways of coping with what defies contemplation. The assumption that *free will* constitutes a dominant factor in the *behavior* of modern societies or nations is incompatible with the evidence for the role of vested interests, such as the military-industrial complex and fossil fuel corporations, in controlling global affairs. Exceptional individuals have always

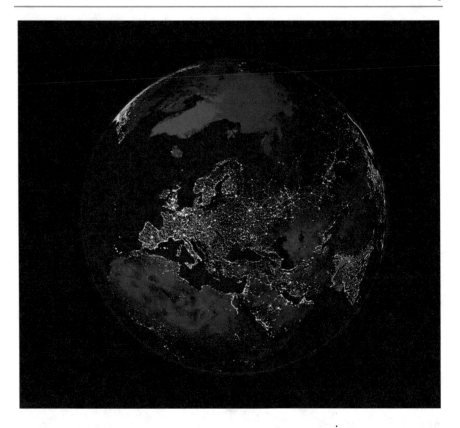

Fig. 12.9 Earth lights, signifying the combustion of fossil fuels around the world—an auto-da-fe of the atmosphere and acidification of the oceans. (NASA)

existed but only in rare instances have their voices prevailed. In the modern world as the quantity of data grows exponentially, not everyone is aware that data is not information, information is not knowledge, knowledge is not understanding, and understanding is not wisdom (Clifford Stoll).

In an evidence-free environment unique ideas proliferate, such as the Anthropic principle, the philosophical suggestion as if a universe cannot exist if there is no conscious witness to observe it, an idea contrary to the geological evidence for the existence of life prior to human observers. Nor is the idea that "Life really could exist in a 2D universe", contrary to everything the geological record and molecular biology are indicating, a plausible one.

And although the planet may not shed a tear for the demise of technological civilization, hope on the scale of the individual is still possible in terms of existentialist philosophy. Going through their black night of the soul, members of the species may be rewarded by the emergence of a conscious dignity devoid of expectations, grateful for the glimpse at the universe for which humans are

Fig. 12.10 Love and the theatre: Limelight—Charlie Chaplin and Claire Bloom

privileged by the fleeting moment: *"Having pushed a boulder up the mountain all day, turning toward the setting sun, we must consider Sisyphus happy"* (Albert Camus, *The Myth of Sisyphus*, 1942) (Fig. 12.12).

Because we believe

Andrea Bocelli

Once in every life

There comes a time

You walk out all alone

And into the night

The moment will not last but then,

We remember it again

When we close our eyes

Like a star across the sky

We were born to die

Fig. 12.11 Love and music: Bob Dylan and Joan Baez

Fig. 12.12 Persephone
supervising Sisyphus in the
Underworld, Attica
black-figure amphora (vase),
c. 530 BC (Creative
Commons)

Eli's Lost World

13

From Seneca's last letter to Nero

Dear Caesar

Keep Burning, raping, killing

But please, please

Spare us your obscene poetry

And ugly music

As nature is fading Eli is resigned to try to live in the moment, mindful of Judith's words: "*It is the transitory nature of things that makes them beautiful. The only things that don't change are dead things. But it is difficult to let go of the idea that things can be permanent—especially when it is the world we are speaking about. The silver lining is that the world seems so much more beautiful when we know that we are saying goodbye to it, at least to the way we know it. Time and time again he returns to the old ideas of Grace and Hope. We hope for a miracle, but we accept the non-appearance of a miracle with grace*" (Fig. 13.1).

Wandering in the desert, in his mind Eli connects with animals around him. The sun-scorched spinifex-clad ledges of the Hamersley Plateau recede into red gorges occupied by waterfalls, reeds and tea tree-fringed ponds, teeming with birds, lizards, goannas, snakes and the odd kangaroo. Eli, perched on a large boulder sheltered from the searing heat, is looking over banded iron formations, the final resting place of a myriad of iron-secreting microbes which two and a half billion years ago lived at the bottom of an ocean. Earth was a different planet then, clouded by an oxygen-starved atmosphere, choked by methane, sulphur dioxide, carbon monoxide and carbon dioxide, over warm and acid seas populated by bacteria, repeatedly inundated by volcanic ash from nearby eruptions and occasional fallout from distant asteroid impacts. Thermal leaching of the iron-rich sediments produced

Fig. 13.1 Burning Earth: electrical lights and fires representing the combustion of fossil fuels around the Earth. NASA. Insert: Eli

layers of blue chrysotile and golden tiger eye fibres which, billions of years later, were to pervade miners' and Eli's lungs:

Sun's splendour now ignites the east

Red burning cliffs, blue fibre mist

That poisons lungs, the air I breathe

Silver needles death bequeath

Of beauty's price, the intertwine

Of mortals glimpse at the divine.

Darkness falls. Eli clears a small patch among the bushes fringing a pond and unrolls his swag.

Darkness fall sparks Venus bright

I lay my weary bones at night

On ancient rocks of Eons past

Beneath the starry sod, a vast

Blanket for my grieving soul

I sleep awake, await a call

In time an orange Luna born

Shrinks to a silver disc, forlorn

In drifting shadows, passing clouds

It fades away through mists, the shrouds

Dim my sense of timelessness

Robs me of a universe

Reflecting, Luna's blinded eyes

That stare at me, Imbrium vies

Toward Nectaris, Cyclops' cry

That haunts my somas vein, I try

Escape my fate, survivals odds

Of self-appointed demi-gods

I gaze into my looking glass

What's meant to be, that too shall pass

Lest I see a human face

In Prometheus' loss of grace

Adam's children, glorious race

Rest in peace in God's solace.

Eli wonders about the next cycle, a question none of his colleagues wishes to reflect upon. He decides to live:

But then emerging from the sea

In space–time rhythm flowing free

Water licks the budding land

Islands born out of sand

Where air-strokes gently ripple waves

Sparks of light each other crave

Wind-swept spores that seed a grove

Sprout all over an alcove

What rhyme or reason springing birth

Upon the ancient face of Earth

Where for a moment you and me

Share the fleeting eternity.

13.1 King Brown

Ascending the brown cliffs (Fig. 13.2a) at dawn Eli pulls himself on to the next ledge, head raised above arm. He freezes. There, a few inches from his stretched hand rests a King Brown snake illuminated by the morning sun, a forked tongue moving in and out (Fig. 13.2b). Eli is paralyzed—the slightest motion and he would be bitten.

Eli utters voicelessly: *I did not mean to disturb* … The snake raises its head: *Humans do not mean what they say, nor do they say what they mean.*

Eli is defensive: We are aware though.

King Brown: *Of self-delusion? Your creation myths decree you are a chosen race while the rest of the animal kingdom is made to serve you, a false superiority underlying your destruction of nature: On the Sixth day God said, Let Us make man in our image, according to our likeness; and let them rule over the fish of the sea and over the birds of the sky and over the cattle and over all the earth, and over every living thing that creeps on the earth.*

Your words fail to discriminate between truth and lie. You use terms couched in mind-altering newspeak pretexts: Open-ended Growth is the philosophy of a cancer cell; the Economy means open-ended self-indulgence requiring the poisoning of land, water and air, radioactive pollution, genetic damage to plants and animals; Security means incinerating the wretched of the Earth, seeding cluster bombs in the fields of poor farmers, killing your young in ritual sacrifices you call war; Freedom becomes anarchy; Democracy leads to the rule of mammon where every dollar has an equal vote.

Your wars in the name of human rights thinly mask a weapons industry seeding the Earth with bombs for profit. You bring your young in front of mind-poisoning commercials, losing a sense of discrimination between truth and lies.

Eli is troubled by the snake's words: There is a price to be paid when a Universe stares at itself through the intelligent eyes of a living species, discovering its physical laws, splitting the atom, decoding the DNA. We are controlled by an evolutionary process whose rationale we do not understand. Have pity on us.

Fig. 13.2 **a** Dales Gorge, Hamersley Ranges; **b** King brown snake

King Brown: *Your comprehension hinges on genetic and anatomical evolution but is blind to the origin and progression of intelligence. You live in flatland. Ancient wisdoms have been replaced by faceless books and one-line tweets. Your ideologies are polarised under a range of guises, left and right, socialist and fascist, symbolizing the struggle between forces which aim to enhance life and those which promote death. You are subject to the tyranny of numbers, counting quantities—money, populations, age—overlooking qualities.*

Eli: *Can you tell me more?*

King Brown: *I cannot. If I knew I would have been forbidden from telling.*

Eli: *Built into science fiction is a plethora of imagined parallel-universe stories, portraying future robotic generations, populating planets, producing synthetic foods, mining asteroids, waging star wars, which serves to divert attention from our planetary plight.*

King Brown: *We snakes have no future in the kind of world you are creating. Having mastered combustion and nuclear fission you exploit, energy orders of magnitude higher than your own respiratory power, destroying my habitat. I really should bite you.*

Eli: *Bite me then. Or tell me what else can I do?*

King Brown: *It was all meant to be this way.*

Eli: *The unthinkable is too scary. How do mass extinctions accord with Darwin's theory of evolution?*

King Brown: *Volcanic eruptions, asteroid impacts and microbial infections can lead to biological catastrophe, but Darwin did not foresee a mass extinction perpetrated by a mammal species. Evolutionary theory revolves around genetic mutation, adaptation and natural selection; it hardly explains the origin of intelligence manifest in the original appearance of the RNA and DNA. It is not clear how purposefulness manifested in all life has emerged.*

Eli: *How did carbon-based molecules evolve to decode the laws of physics inherent in these molecules in the first place?*

King Brown: *You are not meant to decipher these mysteries. A brain cannot fathom its own constitution.*

Eli: *Here we are, teetering on the brink, trying to resolve the fundamental laws of nature, while there is time.*

King Brown: *How much time do we have?*

Eli: *Climate tipping points are looming; nuclear missile fleets are poised on hair-trigger alert.*

King Brown: *I must go back to tend to my young. I will not bite you. You will have to suffer for a while longer.*

The snake disappears down one of the crevices on the ledge.

13.2 The Queen Ant

Traversing through the spinifex for weeks and months, charting the geology, sampling rocks, measuring orientations, taking notes, Eli spends the years, half day-dreaming about some future life out of reach:

Floating this mid-August afternoon

High above the stratocloud

A young and fragile silvery moon

Trails the sun's ignited shroud

Like a suckled lamb

In sprout

Climbing the ridges all these years

Red rocks, the spinifex and crows

Have seen my hopes, my doubts and fears

In ceaseless rhythm ebb and flow　　　　·

In oscillating tides

Of spring

Rising not a flash too soon

Above the eastern vista stars

A reincarnated orange moon

Will once again bewitch my eyes

With other light

Reflect a miracle.

As night falls to the last glow of the setting sun, Eli wanders from his campsite to a tall termite mound, its top smashed by a large dead branch fallen off a nearby gum tree. Through a network of holes of the broken nest a myriad of ants scurry carrying eggs. A group of termites move the huge queen to safety.

Eli reflects: *You are going to last, arthropods are resistant to ionizing radiation. Your proliferation across climate zones will ensure your survival, unlike the otherwise superior mammals and bird species.*

The Queen: *Why do you think the intelligence which governs a termite nest is inferior to that which controls your cities?* (Fig. 13.3).

Eli: *Obviously ...*

The Queen: *We think. How can we design the elaborate architecture of our nests, complete with nurseries, granaries, cooling towers and my royal chamber? We think. Is the design of your refrigerators inherent in your genes or invented by your intelligence? Why don't you give us the credit you give yourself? How do my worker ants coordinate their leaf-stitching, is their language less sophisticated than yours? Why does the Cataglyphis ant's sun-oriented navigational powers not compare with your GPS? Our chemical communication codes are not inferior to your electromagnetic signals.*

Eli: *It is a mystery to me.*

The Queen: *It is the same for other species.*

Fig. 13.3 Termite nest, central Pilbara, Western Australia

How has the language of the bee dance arisen? How does the Queen Bee know she needs to spray anti-fertility pheromones to sterilize her workers, or a damsel fly know it needs to remove a rival male's sperms from a female leg to ensure its own offspring's survival? Does the spider's articulate web design reside in its genes; how does an Archerobie spider plan the theft of prey from a Nephila spider? (Fig. 13.4).

Eli: *How does a Hermit Crab know it needs to place anemones on its back in order to deter an attack by an Octopus? Is this comprehension written into its genes, or does it think?*

The Queen: *To what extent does your understanding of geology originate from your genes, or is it based on your thinking?*

Eli recalls his aunt Salome, a leading geneticist, saying: It is all in the genes, many of which have been identified; he remembers asking: is it then nature or nurture, do we have free will?, with his aunt replying genes are responsible for more than ninety eight percent of your behaviour, and Eli proceeding: we will then have to focus on the remaining two percent?

The Queen: *Genes, computer-like bits and bytes storage units. You know the information stored in a computer is designed by human brains, where do you think the intelligence inherent in the original RNA–DNA biomolecule chains originated from? Where has the vector of purpose inherent in all life originated from and how did it evolve?*

Eli: *There are questions science has not resolved, if it ever can. Humans used to attribute the answers to their gods, or to unknown laws of complexity.*

Fig. 13.4 Natural parallels: **a** a gladiator spider hunting for prey; **b** An Indian fisherman

The Queen: *The evidence is all around you. Inherent in your thinking are assumptions and double standards. How do biomolecules evolve to a brain decoding the laws of nature?*

Eli: *Yet inherent in your colony, as in all other species, is the paradigm of open-ended growth, threatening nature, a principle we humans have inherited.*

The Queen: *We are not as brutal as you, we eat our prey, not torture it; you humans kill for thrill and entertainment in colosseums and homicide shows.*

Eli thinks about the Maya blood-letting cult, the burning of women on the auto-da-fe, the sacrifice of a generation as cannon fodder in the trenches of World War I.

The Queen: *What would nature lose if humans wipe themselves out?*

Eli: *Art and science: Leonardo de Vinci, Shakespeare, Goethe, Dostoevsky, Galileo, Einstein, Martin Luther King, Ghandi, David Attenborough, the list goes on.*

The Queen: *Your species did not listen to their message.*

Silence. The termites carry the Queen into one of the holes, while other ants try and seal the honeycomb openings. Big Tom comes from behind muttering "*silly little ants*".

Eli: *It is an advanced civilization Tom, you ought to beg forgiveness.*

Big Tom prostrates himself in front of the nest, pretending repentance: "*Please pardon my indiscretion Queen ant*".

Eli reflects about the countless insects he inadvertently stepped on over the years.

"*Forgive me too*".

13.3   The Dingo

Eli keeps ascending, accompanied by the five billion year-old yellow star above tracking a fiery arc toward its mid-day zenith, illuminating black ranges slumped like some fossilized monsters on the red sandy plains (Fig. 13.5). The star's rays touch a lone figure carrying a backpack, climbing rocky ledges of ancient lava overgrown by prickly spinifex. Eli, one hundred thousandth the age of the sun, is

Fig. 13.5   The Tomkinson Ranges, central Australia

Fig. 13.6 A rock-carved goanna. Mummawarrawarra Hill, Blackstone Ranges, central Australia

focussed on the rocks ahead, oblivious to neither the unfolding vistas below nor the circling crows above. Humid air and diffuse light herald approaching thunderstorms. Many moons will pass before he is to learn from the Nggaanyatjarra Aborigines that Mummawarrawarra Hill is sacred, a fact which Tommy, his old guide, would not divulge. Perhaps, like many of his brothers, Tommy is too reticent to say "*No*". Perhaps even the fact that this is a sacred site must not be revealed to a descendant of the race which perpetrated the massacres his father still whispers about around the camp fire.

Whatever the reason, ascending the last ledge, Eli reaches a boulder-strewn summit, he collapses exhausted on a small clearing, partly blinded by the nearly climaxed sun. As his eyes adjust to the glare, the outlines of a time-worn rock carving of a goanna imperceptibly begin to materialize on the opposite rock slab (Fig. 13.6), its perched head stretched toward the sun much like his live descendants do. One of them, a small lizard, disturbed by the thud of feet, sneaks under the boulder. The carved goanna remains, defiant, as if saying to the stranger "*How dare you?*".

Eli came across rock carvings before—this one is different though—something compels him to bow his head a little, while his eyes remain mesmerized by the totem. The far horizons vibrate in the mid-day mirage, small distant hills tower above the quivering mirage of imaginary lakes and mountain ranges. Eli feels as if a bone is pointed at him. He remembers—years ago, in a downtown Madras market, a boy sells him a cobra skin. Soon after in high fever, with the last of his cognizance, he lays the sacred object under a rock to appease the angry spirits.

This time there is no turning back. The climaxed sun radiates brain-piercing beams; salty perspiration drips down Eli's forehead and burns his eyes with sparkling crystals of light. Eli dozes off listlessly, the dancing hills recede in the distance, the crows circle ever lower. He wakes up to a heavy flutter of wings, jolted by a blunt thump and searing white hot pain. The sun's fireball in the north explodes into a myriad of shattered crystals.

Is it a starless night or am I dreaming? Tell-tale sounds of the night resonate, the swish of tails, an eerie howl, soft whispering voices, and a muffled echo of a didgeridoo.

"*It is us*", the smiling broad black faces of Nugget, Tommy and Jim materialize out of the shadows, thick lips murmur:

"*We're sent by the Great Serpent, he grants you a last wish—you see, we made a plea on your behalf*".

Silence. I despair, all my life I longed for a wish, now my mouth opens and closes with great effort, voicelessly. The elders read my mind,

"*Why a world, what is life, is there a creator, who created the creator, is there free will?*"

The elders sound alarmed,

"*No one but the Great serpent himself is allowed this knowledge, you must find out for yourself, before the next sunrise*".

Reverberating voices recede, "*Before sunrise ... before sunrise*".

My hands touch and feel where my eyes used to be. The pain has gone. For what feels like eternity I remain lying in total darkness.

Twinkles of light tingle my eyes. An irresistible force catapults me skyward ... now it lets go ... I spin in free fall, flapping my wings I shore up to be propelled forward at the speed of light. I am a brain bubble, a photon caught in a snaking space–time tunnel, spanning shooting star galleries and red nebulae lit by sparkling crystal windows. Dim figures come into view through the windows, calling me, cajoling me, luring me with colour and sound. Four sirens dance out of one of the windows, shaking cymbals and singing:

"*We are adenine, cytosine, guanine and thymine,*

Our joined hands the goddess DiaNA define,

Mystery life strings out of us intertwine,

Multiply to the end—decrees the divine,

Swallow or die reins the futile design.

We hope you won't mind a fate so malign".

The sisters swing toward a looming star which gradually acquires the shape of a shrine. Strapped on an altar is a tearful blue eyed girl wearing a brown and green matted skirt. Behind her towers a white-hooded priest wearing a GNP label and holding a dollar sign-handled dagger, roaring:

"Confess Gaea, reveal your treasure,

Let countless billions live in leisure,

For you are but a quarry pit,

A corridor to heaven in transit,

Your green lovers in hell shall burn,

The next sun rise will be your turn".

"Help!" I fly screaming toward a window shaped like a TV screen, framing a human figure.

"No need to be excessively emotional", scoffs a rational economist to the background of canned laughter, he adds:

"Gaea's shares of late abated,

The market force cannot be sated,

trading in Mars futures has been great,

The planets are waiting, abandon the Earth".

He is interrupted by a horrendous shriek sounding like the crush of 100 space crafts and a talking head announcing a news flash:

"We are your XYZ.000 NEWS station

The ultimate voice of the world's top nation

Proud to proclaim that we are best,

The All Ordinaries shot up, forget all the rest,

Floods caused 245 million dollars damage in Maine,
   (His voice quivers as he reads *"dollars"*).

A gunman killed three people today in Balmain,

The Broncos won the grand final in spite of the rain".

A bunch of business executives, laughing all the way to the bank and waving.

"$ Futures Mission $"* and *"$ Value-Added Vision $" posters, sing:

"By the year two thousand, by the year two thousand"... *"Spend, save, spend, save, spend, save ..."*

Their voices thump in my brain like jack hammers; I try to ask whether there is any news from the Amazon forest, the Ukrainian soils, the Arctic Ocean, Irian Jaya or the ionosphere, when a huge asteroid whirls by in a flare.

"Ja wohl!" thunders Lucifer, standing on one leg in a crater:

"Black holes devour galaxies over a span,

Humans unmake ozone plants patiently spun,

Entropy grinds down every grand plan."

Darwin's white bearded face gestures mockingly from a passing red giant:

"Monkey sees—monkey do,

Natural selection is the one vital clue,

Where life is concerned there is no taboo".

"*No!*" I cry voicelessly, "*No!*", there must be a reason, piano playing monkeys can't reproduce a Mozart concerto at random, computers are not synthesized from metal in sea water accidentally.

"Don't bet on it my boy, they will do so as long as the clients pay" says Lucifer.

I try to ask Darwin: *"is there anything that is impossible, anything humans can't do?"* In response, a blood curdling murmur rises from a line of naked figures with shaved heads, moving toward a compound with smoking chimneys. To the tune of Wagner's Twilight of the Gods, Lucifer, this time wearing a swastika arm band, piercing hypnotic eyes overlooking a small moustache, proclaims:

"Ja wohl, ha, ha, morals are for the faint hearted, read Nietzche!"

"Your methods fall short", exudes Doctor Strangelove from the cockpit of a passing star wars ship:

"We have beaten you to the patent".

As the sun's rays approach Gaea's chest the GNP-labelled priest raises his dagger, DiaNA's daughters form rings around the nuclear shuttle, replicating. From the distance the double helix can be hardly distinguished from the dagger's dollar sign. DiaNA's dancing rings look more and more like an atom girdled by electrons —the circle has come a full turn.

The short-lived winter twilight fades beyond the Blackstone Ranges, the rising moon's orb silhouettes dark clouds in the west, drawing a lament from a dingo (Fig. 13.7) sniffing a shivering body. She walks away, returns and lies by the blind man's feet.

I see a spark of light; I glide back to Earth effortlessly over azure blue oceans, broken volcanic peaks surrounded by coral reefs and lagoons. I flap my wings and land into a lagoon like a pelican. Small waves die on the shore one by one, weary from their long journey across the oceans—each foaming in its fading spasms, collapses, crawls up the beach, to be pulled back to its mother's womb.

A group of Polynesian sailors and flower clad girls approach me:

"What is your name?" I ask one of them.

"*RiNA*", she replies, a warm glow in her voice. I sigh—I always longed for you, I always run away from you, you grow with child, multiply, fulfilling DiaNA's decree. I soar into the south Pacific sun escaping the gravity pull of the girl's pleas:

"Eli come back, come home!".

Fig. 13.7 A rock carving of a dingo. Blackstone Ranges, central Australia

I approach the sun, my wings melt, and I plunge back into the white surf, an Icarus, at last resting among corals, anemones and variegated fish. A dolphin caresses me with its fins:

"Sorry my friend, I can't tell you the answers - try music".

"Try mathematics",

Says the originator of relativity, his white mane reflected in the ripples above the sea bed on which his ashes have been scattered:

"God is too kind to play dice,

The universe is not a chance,

My brain's insight proves my stance".

I want to ask what are the limits to human perception, the white mane dissolves in the water.

"Space/time is but a whole,

Probability's claim, the final toll",

Says a quantum physicist, puffing on a cigar.

"I don't follow chaos", I wonder: *`how come a butterfly's wing flutters trigger a hurricane?"*

From an underwater stream an astronomer looks at me reflectively:

"Starstuff pondering the stars,

Clusters of atoms considering the evolution of atoms".

I want to tell him about the Great Serpent, to no avail. I return to stroke the dolphin, its skin feels warm, furry, like that of a dog—great God, there is a big dog lying on top of me!

The dingo, sensing the quiver of Eli's body, licks the blood from the sick man's face. A muffled hum draws closer, soon big rain drops start splashing from the jet black sky. The dingo is busy dragging some branches, returns and lies on Eli, protecting him from the worst of the drenching rain. Only the stone carved goanna remains perched skyward, as it has always done. At the base camp Eli's party members wonder why he did not turn up on the evening radio schedule.

"It's probably the thunderstorms, the static", says John. It's not the first time Eli went walkabout, returning safe a few days later. With the now steady downpour the ground will be too wet for a search party to be sent for some time.

My hand keeps stroking my companion—not without a sense of guilt. As if I did not know about the hysteria, the witch hunts and massacre of dingoes in the wake of the events at Uluru? The dog should have left me yet I am receiving her protection under false pretences. My other hand senses the gaping holes that once were my eyes, my inflamed forehead, and my chaotic pulse. I am dying. Didn't Nugget, Tommy and Jim say I was granted a wish? The dingo has fallen asleep, sporadically uttering soft sounds. She may be dreaming, perhaps about her pup she left somewhere, perhaps about rabbits. Despite the cover I am drenched to the bone; I want to dig into a dry place. I keep scratching the ground with my fingernails—a huge chamber opens.

At the centre of the cavern there is a high throne on which sits a crowned ant wearing brilliant red and green armour, surrounded by nursing maids and body guards. She looks at me with disdain through a pair of glasses:

"I was expecting you, you butcher of thousands of my folk".

Me? All the ants I have been stepping on over the years, never giving it a thought.

"As if we have not a mind, not a soul",

The Queen reads my thoughts accurately, she adds:

"As if we can't match your social communications, organization, engineering, warfare".

You are such tiny, genetically programmed automatons, I reflect.

The Queen fumes:

"What has size to do with it? How big are DiaNA's molecules, your neurons, your microchips? We may be small, we don't suffer from your kind of insanity though, your craze for gold".

I see images of slaves in Athens' silver mines, Spaniard soldiers torturing Inca Indians, Bantu miners drilling deep under the Witwatersrand goldfields. I want to ask—is there is any atrocity humans will not commit?

"There are no limits, what makes you think you're not merely an instrument of Lucifer's will?"

Furiously the Queen sends a detachment of guards towards me.

I wake up stung by green ants escaping the rain and attracted to the warmth of my body. I itch all over, I am thirsty, my hand is sticky, I lick it, it tastes like milk—I grab the dingo's milking bud with my mouth. The rich life-giving liquid pervades my body. I smile to myself—I may found the new Rome like Remus and Romulus; I am unaware that, through suckling me I became a part of the dog's litter—the dingo is allowing me into her clan.

This knowledge is not lost on the Great Serpent. The huge red-eyed snake, curled in a granite cave, tosses and turns during a bad night's sleep. The white man has trespassed the sacred goanna, he has already been punished, he can be granted the customary last wish as he is asking for a deeper knowledge than allowed, yet now he is protected by the Great Dingo.

Large holes are torn in the cloud cover, revealing a myriad twinkling stars. In the east the black melts into a dim halo. Spinifex pigeons rattle under nearby bushes. The gurgling of a kookaburra interrupts the chirping of honeybirds. Now the morning star looms brightly, Eli's deadline is nearly expired; the Great Serpent decides to summon him to the cave.

The sleeping dingo's body rises and falls rhythmically. The rich flavour of her milk makes me feel like an infant, I am pervaded with a sense of belonging I never had before. Once an outcast boy in a war torn homeland, I am immersed in a golden ocean of tall wheat stems flooded by the sun's yellow. The sun takes forever plotting its daily course, pervading me with immortal bliss, safe and at one with the Earth. The knowledge dawns on me: I am unmarked by any clan, Jung's sacrificial goat lurks inside, the Pan goat, Dionysus' life-loving symbol sacrificed in redemption. I reflect in my daze—there is no escape now.

"There is no escape now",

Echoes the Great Serpent (Fig. 13.8), raising its huge python head:

"We are not as vindictive as your Gods are, though, you have received good references".

It nods toward the Great Dingo, busy licking a bone in the corner of the cave.

"My last wish" I utter a sound, amazed my speech has come back.

The Great Serpent looks almost sad:

"You will not understand, your questions are formulated in words. Nature can be kind or cruel, only words can lie. Remember Babylon's tower, remember Orwell's Newspeak?.

I think I know what the old gentleman is getting at. I think about a French Nobel laureate gagging himself in protest amidst his colleagues in a pre-world war III conference.

The Great Serpent adds: *"Your brain has long denied the intelligence of indigenous people; you still deny the soul of living creatures, of trees"*.

The Great Dingo raises its head from the corner, releasing a ghostly howl.

Fig. 13.8 Rock carving of snakes, Reynold range, central Australia

I try and ask, *"Has it all been meant to be like this, has anyone ever had a choice in the matter, free will?"*.

"Gaea has her checks and balances—earthquakes, floods, asteroid impacts, epidemics, technical civilizations—life has never been easy, not even in the Dreamtime",

The Great Serpent sighs. I feel for the mottled old creature, its lacklustre scales peeling off. I keep on nagging with some guilt:

By accident or by design?

"Every accident is a part of a plan", answers the Great Serpent,

"Foolishly we gave you a wish, not realizing what you would choose to ask for. I am tired, I have seen too much ..."

The Great Serpent curls back into the darkness of the cave, toe the end of time (Fig. 13.9), his enormous tail crushing the floor, raising a cloud of dust. Loud echoes reverberate through hidden subterranean caverns; they subside and fuse into the sweet sound of a pan flute. The music inundates me with an immense sensation of joy such as I never knew before. In a splash the sunshine bursts into the cave along with dancing centaurs and nymphs led by a horn playing black goat-headed minotaur. They embrace me:

"Welcome to the Gaea family".

I reciprocate

Fig. 13.9 *Time*. A natural mortar and pestle, Komati River, Swaziland: Signifying the passage of *time*

The dingo, awoken by Eli's sudden hug, rises from the blind man she tunes her ears to the remote hum of an engine. With his tracking skills, Tommy managed to trace Eli's route despite the washouts and lead the search party to Mummawarrawarra Hill. They approach, seeing a lone dingo running in the opposite direction. A gun blast …

Bibliography
Peer Review Papers and Books

Afflerbach H (2014) The soldiers across Europe who were excited about World War I. The Conversation. August 4, 2014 3.23pm AEST

Anderson K (2015) Is the IPCC overly optimistic on our climate? Policy@Manchester Blogs. http://blog.policy.manchester.ac.uk/posts/2015/10/is-the-ipcc-overly-optimisticon-our-climate/

Ananthaswamy A, Douglas K (1998) New scientist. https://www.newscientist.com/article/mg23831740-400-the-origins-of-sexismhow-men-came-to-rule-12000-years-ago/

Arendt H (1963) Eichmann in Jerusalem. New Yorker. https://www.newyorker.com/magazine/1963/02/16/eichmann-in-jerusalem-i

Avramik SM (1992) The oldest records of photosynthesis. Photosynth Res 33:75–89

Bonabeau E (1999) From natural to artificial swarm intelligence. https://dl.acm.org/doi/book/10.5555/554879

Brookfield ME (2010) The rise of Egyptian civilization. In: Landscapes and societies, pp 91–108

Buis A (2020) Milankovitch (orbital) cycles and their role in earth's climate. NASA Global Climate Change

Burchett W (2015) Honest history: Wilfred Burchett in Hiroshima: highlights reel. https://honesthistory.net.au/wp/wilfred-burchett-in-hiroshima-highlights-reel/

Burke KD et al (2018) Pliocene and Eocene provide best analogs for near-future climates. Proc Nat Acad Sci 115(52):13288–13293

Buschinger A (1986) Evolution of social parasitism in ants. Trends Ecol Evolut 1:155–160

Callaghan ML (2016) Bird brains are richer in neurons than mammal brains. Popular Science, June 14, 2016

Carrigan HL (2014) Unbearable pain: faith after the holocaust: Stuart Matlins. https://www.publishersweekly.com/pw/home/index.html

Camus A (1942) The Myth of Sisyphus. Penguin Books, 208 pp.

Chomsky N, Herman ES (1995) Manufacturing consent: the political economy of the mass media

Chomsky N (2010) Human intelligence and the environment. Int4r Social Rev Issue #76

Cullen HM et al (2000) Climate change and the collapse of the Akkadian empire: evidence from the deep sea. Geology 28:379–382

Dahlman L, Lindsey R (2020) Climate change: ocean heat content. NASA. https://www.climate.gov/news-features/understandingclimate/climate-change-ocean-heat-content

Dawkins R (2006) The god delusion. Penguin Books. https://www.penguin.com.au/books/the-god-delusion-9781784161927

de Vallombreuse P (2015) Galerie Argentic. https://wsimag.com/art/17818-pierre-de-vallombreuse-souveraines

deMenocal PB (2001) Cultural responses to climate change during the late Holocene. Science 292:667–673

Diamond J (1911) How societies choose to fail or succeed. Viking Press, Penguin Random House, USA, p 592

Ellis GFR (2005) Physics, complexity and causality. Nature 435:743

Ellis GFR (2012) Top-down causation and emergence: some comments on mechanisms. Interface Focus 2(1):126–140

Emery NJ (2006) Cognitive ornithology: the evolution of avian intelligence. Philos Trans R Soc Lond B Biol Sci 361(1465):23–43

Engelbrecht HC, Hanighen FC (1934) Merchants of death: a study of the international armament industry. Taylor and Francis, 338 pp.

Fantham E et al (1994) Women in the classical world. Oxford University Press, Oxford

Foreman A (2014) The Amazon women: is there any truth behind the myth? Smithsonian Magazines

Gillooly JF (2010) Eusocial insects as superorganisms. Commun. Integr. Biol. 3(4):360–362

Glikson AY (2013) Existential risks to our planetary life-support systems. The Conversation, September 2013

Glikson AY (2016) Global heating and the dilemma of climate scientists. ABC The Drum. 29 Jan

Glikson AY (2018) The methane time bomb. Energy Procedia 146:23–29

Guesco M (2017) The origins of slavery. Oxford Bibliographies. https://www.oxfordbibliographies.com/

Halpern DF (2012) Sex differences in cognitive ability. Taylor and Francis, 445 pp.

Hansen J et al (2008) Target atmospheric CO_2: where should humanity aim? Open Atmos Sci J 2:217–231

Hansen J (2016) Can't burn all fossil fuels without creating a different planet. Citizens' Climate Lobby

Huxley A (1932) Brave new world. New Longman Literature, 311 pp.

Jenner L (2020) NASA aids disaster response after Eta and Iota Hit Central America. NASA Hurricane And Typhoon Updates

Jensen D (2016) The myth of human supremacy. Penguin Random House

Koestler A (1978) Janus: a summing up. Amazon, 368 pp.

Kompanichenko VN (2000) Average lifetime of intelligent civilization estimated on its global cycle. A new era in bio-astronomy ASP Conference Series 213

Kneller TL (1993) Role of women in Nubia. University Pennsylvania African Study Centre. https://www.africa.upenn.edu/Articles_Gen/Role_Women.html

Knight D (2020) Women rule: a look at 5 matriarchal societies throughout history. https://www.mydomaine.com/matriarchal-societies

Krajick K (2019) Wallace Broecker, Prophet of climate change. State of the Planet, Earth Institute, Columbia University. https://blogs.ei.columbia.edu/2019/02/19/wallace-broecker-early-prophet-ofclimate-change/

Langergrabe KE et al (2013) Male–female socio-spatial relationships and reproduction in wild chimpanzees. Behav Ecol Sociobiol 67:861–873

Lewis S (1935) It cannot happened here. Penguin Books, 400 pp.

Lindsey R (2018) Research cruises reveal global warming reaching the deep Southern Ocean. NOAA Climate.gov

Marais EN (1925) The soul of the white ant. https://www.goodreads.com/book/show/1419087. The_Soul_Of_The_White_Ant

Mayor A (2014) The Amazons: lives and legends of warrior women across the ancient world. Princeton University Press

Maguire D (2018) Australian birds `Firehawks' deliberately spread fires in incredible hunting technique

Nofil B (1953) The CIA's appalling human experiments with mind control. https://www.history.com/mkultra-operation-midnight-climax-cia-lsdexperiments

Nyman P (1994) Methane vs. carbon dioxide: a greenhouse gas showdown. One green planet. https://www.onegreenplanet.org/animalsandnature/methanevs-carbon-dioxide-a-greenhouse-gas-showdown/

Ogburn SP (2013) Ice-free Arctic in Pliocene, LAST TIME CO_2 levels above 400 PPM: sediment cores from an undisturbed Siberian lake reveal a warmer, wetter Arctic. Nature magazine, May 10, 2013

Olmos D (2019) When watching others in pain, women's brains show more empathy. UCLA Newsroom. https://newsroom.ucla.edu/stories/womens-brainsshow-more-empathy

Oreskes N, Conway EM (2015) Merchants of doubt: how a handful of scientists obscured the truth on issues from tobacco smoke to climate change: Amazon.com

Orwell G (1949) Penguin Books 336 pp.

Powell CS (2017) How humans might outlive Earth, the sun and even the universe. https://www.nbcnews.com/science

Primo L, Woolf S (1987) If this is a man/the truce: Hachette essentials. Auschwitz Trilogy #1-2. Amazon.com.au. 453 pp.

Ramamurthy R (2015) The rise of the collective ego. https://www.linkedin.com/pulse/rise-collective-ego-ravi-ramamurthy/

Raymond C (2020) The emergence of heat and humidity too severe for human tolerance. Sci Adv 6:19

Remarque EM (1929) All quiet on the western front. Mass Market Paperback, 296 pp.

Rettner R (2010) Insect colonies function like superorganisms. Live science. https://www.livescience.com/8020-insect-colonies-functionsuperorganisms.html

Reynolds H (2013) Forgotten war. Booktopia, 288 pp.

Richardson JH (2018) When the end of human civilization is your day job. Esquire. https://www.esquire.com/news-politics/a36228/ballad-of-the-sadclimatologists-0815/

Roberts EM et al (2016) Oligocene termite nests with in situ fungus gardens from the Rukwa Rift Basin, Tanzania, support a Paleogene African origin for insect agriculture. Plus One. https://doi.org/10.1371/journal.pone.0156847

Schellnhuber HJ (2009) Tipping elements in the earth system. Proc Nat Acad Sci 106(49):20561–20563

Schmidt AR et al (2012) Arthropods in amber from the Triassic Period. Proc Natl Acad Sci U S A 109(37):14796–14801

Sherefkin J (2016) Immortality and the fear of death. New York Public Library. https://www.nypl.org/blog/2016/02/04/immortality-fear-death

Stanley M (1987) Periodic mass extinctions of the earth's species. Bull Am Acad Arts Sci 40:29–48

Slater S (2014) Monsters of the mind: is there a perceptual basis for the darkness that lurks within? Psychology Today. https://www.psychologytoday.com/intl/blog/the-dolphindivide/201402/monsters-the-mind

Sleight J (2019) Scientists and the bomb: 'the Destroyer of Worlds'. Global Zero. https://www.globalzero.org/updates/scientists-and-the-bomb-the-destroyer-ofworlds/

Stapledon O (1930) Last and first man. Amazon CA, 246 pp.

Steffen W et al (2007) The anthropocene: are humans now overwhelming the great forces of nature. AMBIO J Human Environ 36(8):614–21

Steffensen JP (2008) High-resolution greenland ice core data show abrupt climate change happens in few years. Science 5889:680–684

Taylor S (2014) The psychology of war: why do humans find it so difficult to live in peace? Psychology Today

Turner JS (2011) Termites as models of swarm cognition. Swarm Intell 5:19–43

Vaughan A (2020) Fracking wells in the US are leaking loads of planet-warming methane. https://www.newscientist.com/article/2241347-fracking-wells-in-theus-are-leaking-loads-of-planet-warming-methane/

Vince G (2014) The new superorganism taking over the Earth. BBC. https://www.bbc.com/future/article/20140701-the-superorganism-engulfingearth

Wallace-Wells D (2017) The uninhabitable earth. Penguin, 336 pp. https://doi.org/10.1002/jmor.1050220206

Ward L (2001) Poor give more generously than the rich. https://www.theguardian.com/society/2001/dec/21/voluntarysector.fundraising

Weiss H et al (1993) The genesis and collapse of third millennium north Mesopotamian civilization. Science 261:995–1004

Wells HG (1898) War of the world. https://theconversation.com/guide-to-theclassics-the-war-of-the-worlds-128453

Wheeler MW (1911) The ant-colony as an organism. J Morphol

Wrangham R, Peterson D (1997) Demonic males: Apes and the origins of human violence. Mariner Boos, Boston, 350 pp. https://doi.org/10.1002/jmor.1050220206

Index

Printed in the United States
by Baker & Taylor Publisher Services